INCREDIBLE ELEMENTS

A TOTALLY NON-SCARY GUIDE TO CHEMISTRY AND WHY IT MATTERS

METRO BOOKS
New York

An Imprint of Sterling Publishing Co., Inc.
1166 Avenue of the Americas
New York, NY 10036

METRO BOOKS and the distinctive Metro Books logo
are registered trademarks of Sterling Publishing Co., Inc.

ISBN 978-1-4351-6468-0

For information about custom editions, special sales, and premium and corporate purchases,
please contact Sterling Special Sales at 800-805-5489 or specialsales@sterlingpublishing.com.

Manufactured in China

2   4   6   8   10   9   7   5   3   1

www.sterlingpublishing.com

Credits: Interior layouts by Tony Seddon

Image credits

6 © Alamy | Nature Picture Library, 7 (top), 48, 89, 110 Wellcome Library, London, 12 © Alamy |
North Wind Picture Archives, 14 © Shutterstock | S_Photo, 15 © Creative Commons |
klimaundmensch.de, 16 © Shutterstock | Albert Russ, 17 © Creative Commons | Henry Walters, 19
© Creative Commons | Arkadiy Etumyan, 21 (top) © Shutterstock | Andrei Ch, 21 (bottom) ©
Shutterstock | italianvideophotoagency, 22 © Shutterstock | Aleoks, 38 © Shutterstock | Dan Logan,
43 (top) © Getty Images | Heritage Images, 43 (bottom, left) © Creative Commons | Jon Zander, 43
(bottom, right) © Creative Commons | Daniel Grohman, 79 © Shutterstock | Vadim Petrakov, 80 ©
Shutterstock | Denis Burdin, 83 © Shutterstock | worraset, 87 © Creative Commons | Rama, 91 ©
Getty Images | Clive Street, 109 © Creative Commons | Luigi Chiesa, 113 © Shutterstock |
Morphart Creation, 123 © mmmdirt, 135 © Creative Commons | Rob Lavinsky, iRocks.com, 138 ©
Shutterstock | Christian Draghici, 139 © Shutterstock | hddigital

INCREDIBLE ELEMENTS

A TOTALLY NON-SCARY GUIDE TO CHEMISTRY AND WHY IT MATTERS

Joel Levy

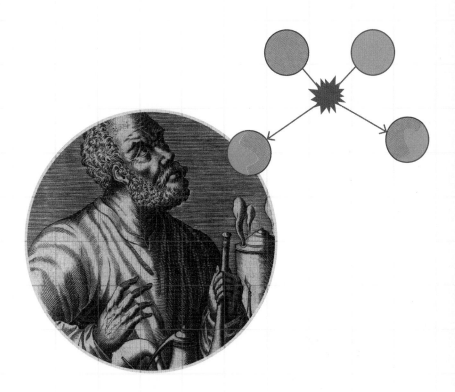

METRO BOOKS

NEW YORK

Contents

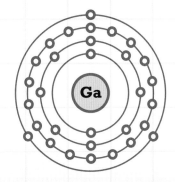

Introducing Chemistry

Technically speaking, chemistry is the study of the elements and the compounds they form, but in its broader sense it is so much more.... It's the science of everyday life, of the matter that makes up the world, and every single one of us can transform it in ways that seemed magical to our ancestors, and can still seem fantastic today.

An Inspiring and Transforming Science

For many of you, the word "chemistry" will summon up images of test tubes and Bunsen burners, white lab coats, strange smells, and the vague hope of an explosion. Yet this brief school-days experience of the subject does it little justice, as this book sets out to prove. In these pages you'll learn how chemistry transformed mankind,

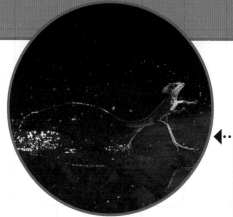

A **Jesus lizard** (*Basiliscus vittatus*) running on water.

"Chemistry ... offers one of the most powerful means towards the attainment of a higher mental cultivation ... because it furnishes us with insight into those wonders of creation which immediately surround us."

—Justus von Liebig

gave life to civilization, sparked the imagination of mystics and magicians, and inspired the greatest minds in history.

You won't need any prior knowledge of chemistry, as everything from the most basic concepts to the most profound laws of matter will be explained in a clear, concise way. Just bring a sense of adventure and wonder. Along the way you'll meet some strange and incredible characters and learn a host of fascinating trivia, from why bread rises and ice floats to how a lizard can walk, or at least run, on water.

Uncovering the Mystery of Matter

Chemical knowledge is growing all the time, faster than most people imagine. More than 8 million different chemical substances, both natural and artificial, are now known. As recently as 1965 only 500,000 had been identified and produced, and even this far outstripped the wildest imaginings of the chemists of 200 years ago.

To avoid being overwhelmed by the enormous scope of chemistry, this book focuses on inorganic chemistry—the branch that deals with all the elements (see p. 9) other than carbon, and with their compounds. Some simple carbon compounds, such as carbon dioxide and calcium carbonate (chalk or limestone), also come under the study of inorganic chemistry.

Most of the history of chemistry relates to inorganic chemistry, so their stories are largely intertwined—and what a story it is. One of the greatest

One of the greatest adventures in the history of ideas, the development of chemistry is a tale of obsession, greed, danger, hope, and inspiration.

Alchemy, an early precursor to chemistry, was a strange hybrid of science and art, based on early Greek philosophical ideas. Alchemists were highly secretive about their work.

In the 17th century a new scientific approach to chemistry emerged, personified in the life and career of **Robert Boyle**, who came to be known as the "father of scientific chemistry."

A key turning point for chemistry was in 1869, when Russian chemist **Dmitry Mendeleyev** established periodic law and devised the first periodic table of elements.

the ancient Egyptians' sophisticated chemistry, and the ancient Greeks' attempts to explain nature. It looks at the mix of science and magic known as alchemy, the transforming effects of the scientific revolution, and the search for the elements. Building to a climax with the founding of periodic law, the key to all chemistry, the book ends with the fulfillment of an age-old quest: elements that change into other elements.

adventures in the history of ideas, the development of chemistry is a tale of obsession, greed, danger, hope, and inspiration, and each chapter of this book relates a chapter in that tale.

Following the millennia-long quest to uncover the secrets of matter, this book charts humans' early use of chemistry in fire-making and cooking,

At every stage in this epic journey, essential concepts of chemistry are introduced and explained with a minimum of math and formulae so that you too can decipher the hidden language of the elements that helps make sense of the world around us.

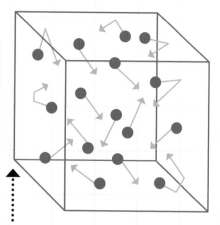

Copper (Cu) displacing silver (Ag) from a **silver nitrate** solution.

Gas particles move about randomly, in straight lines, until they collide with the solid walls of their container.

ATOMS, MOLECULES, ELEMENTS, AND COMPOUNDS

Before going any further, let's look at some very basic terms and concepts. Two pairs of terms that chemists use a lot are "atoms" and "molecules," and "elements" and "compounds." What's the difference between them?

An **atom** is the smallest unit of any substance.

An **element** is the purest form of matter, with no other ingredients mixed in. It is made up of atoms that are identical to each other, and they are different from the atoms of all other elements. For instance, the element carbon contains only carbon atoms. The atoms of some elements, such as helium, stay as separate atoms.

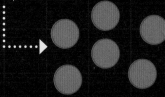

The atoms of other elements are chemically joined together to make **molecules**. For example, in its pure form oxygen is a gas made up of molecules, each of which is composed of two oxygen atoms bonded to one another. ....➤

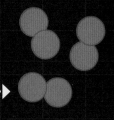

Currently there are 118 known elements.

A **compound** contains two or more different elements that are chemically bonded. Compounds are formed during chemical reactions, where atoms and molecules interact and adopt new combinations and arrangements. An example of this is when carbon and oxygen combine to form the compound carbon dioxide: carbon atoms react with oxygen molecules to form carbon dioxide molecules.

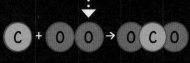

All of these terms are explained in more detail later in the book (for atoms, see pp. 28-29; for elements, see pp. 24-25; for chemical bonds, see pp. 64-65; and for chemical reactions, see pp. 38-39).

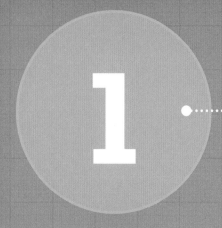

1

Chemistry in the Ancient World

This chapter explores the rich chemical heritage of antiquity and beyond, reaching back in time to the prehistoric era, ancient Egypt, and ancient Greece. Along the way it introduces some of the important concepts of matter and energy—not to mention the secret of good toast. Prehistory and ancient history may predate science, but they were eras of great technological sophistication, including surprisingly advanced chemical technologies. They also witnessed the birth of modern ideas of matter and the elements.

Prehistoric Chemistry

Chemistry might seem to be a modern science. In fact it was seen as *the* science during the Enlightenment, an 18th-century European cultural movement that believed science and logic gave people more knowledge than tradition and religion. But actually chemistry is as old as human culture—you could say it was our use of chemistry that made us human. Whether they realized it or not, our ancestors dating back to the dawn of humankind were using the basic principles of chemistry.

Fire Starters

One of the turning points in human evolution was when our ancestors first began to control their environment using combustion. "Combustion" means the oxidation of carbon: the bonding of carbon with oxygen in an exothermic chemical reaction (one that releases energy in the form of light and heat)—in other words, fire. The conditions for spontaneous combustion of carbon are very rare on Earth because typical combustion reactions need activation energy—an energy boost that gets the reaction started (see pp. 38–39).

There's evidence that *Homo erectus*, an ancestor of modern humans (*Homo sapiens*), used fire to clear habitats and probably to flush out game. Possibly the earliest fire-users took advantage of naturally occurring fires started by lightning strikes. But by the time *Homo sapiens* had evolved, if not before, our ancestors had learned how to produce enough activation energy to start combustion for themselves—either by striking sparks from flints or by rubbing bits of wood together to generate heat by friction.

A host of other technological advances followed on from this. First, and perhaps most important for human evolution, was the discovery of the chemistry of cooking (see pp. 14–15). This made a much wider range of food edible and opened up new diet options with high calorie and protein content.

Prehistoric man's **discovery of how to make fire** was a pivotal step in human evolution.

Paintings made with red and yellow ocher pigments have been found in Stone Age sites in Europe, including this bison image (ca. 15,000–16,500 BCE) from the cave of Altamira in Spain.

Pyrotechnic Skills

Mastering the chemistry of combustion led to many other pyrotechnic technologies—"pyrotechnics" stems from the Greek words *pyr* ("fire") and *tekhnē* ("art"). One of the earliest was the treatment of the pigment ocher. Ocher is a type of clay tinted yellow-brown with the iron ore hematite—an iron compound known as hydrated iron(III) oxide, which has the chemical formula Fe_2O_3 (see pp. 110–111 for an introduction to chemical notation). Prehistoric man discovered that heating ocher at 500–540°F (260–280°C) caused a chemical reaction called calcination, which produced a still wider range of colors—in particular a striking red.

After treating ocher the pyrotechnic artists may have turned to flint and clay. Heat treatment changed the structure of flint to produce sharper edges so better tools, while firing clay led to the craft of pottery.

THE METAL AGES

The chemistry of pyrotechnology made possible the development of metallurgy (the study of metals) and the development of prehistoric civilization from the Stone Age, to the Copper Age, to the Bronze Age, and finally to the Iron Age. The sequence of these ages is explained by the chemical properties of the metals. Metals that react easily with oxygen and other elements are never found in a pure form in nature, while those that don't react with other elements are found in pure forms, and are easier to mine and work.

Gold is the least reactive metal and was probably the first worked by humans, but its softness meant it had little use beyond ornament. Copper is also found in its pure form and pyrotechnology allowed ores to be smelted and extracted metals to be melted and cast (poured into a mold). Ores of copper and tin are sometimes found together, and when smelted they would have produced an alloy (a mix of metals), in this case bronze. Iron ores are more abundant than those of copper and tin, but iron has a much higher melting point, making it hard to smelt (extract the metal from the ore) until kiln technology had advanced. By around 1100 BCE, however, ancient metallurgists discovered that reheating impure iron with charcoal produced steel.

The Chemistry of Cooking

If you cook then you are a chemist, with the kitchen as your laboratory, because cooking *is* chemistry. When we cook, we use heat to produce chemical reactions between food molecules, changing them into other molecules. This gives the cooked food new properties, changing its flavor, smell, color, consistency, nutritiousness, and even toxicity.

When food is cooked, its chemistry changes as a result of heat causing many of the complex molecules to break down into smaller molecules. Some of these changes are important for human consumption of food, as they allow us to digest the food chemicals more successfully— for instance, tough proteins in meat are broken down into more digestible forms during cooking. Heat also helps reactions to occur between food chemicals.

Have you ever wondered why **a freshly baked loaf** of bread looks, smells, and tastes so delicious? It's all thanks to the Maillard reaction.

Yeast from the East

Not all cooking involves heat. Yeast's remarkable chemical abilities were probably first used by prehistoric man, but the first records of brewing come from Mesopotamia, an ancient region

The French chemist **Louis Camille Maillard.**

in the eastern Mediterranean, ca. 4000 BCE. Beer was also brewed in ancient Egypt around the same time.

The kneading of dough in bread-making is an example of mechanical alteration of food chemistry. Kneading causes proteins in the dough to form long, elastic chains, which allow the dough to rise by trapping the gas produced by the action of yeast. Yeast is a fungus that converts sugars into ethanol, a type of alcohol, giving off carbon dioxide in the process—a reaction known as fermentation. During baking, the alcohol is mostly driven off by heat, but fermentation as an end in itself is the basis of brewing.

Stovetop Magic

One of the most important cooking reactions was discovered by the French chemist Louis Camille Maillard (1878–1936). In 1912 he found that if enough heat is applied, then proteins and carbohydrates will react together to produce a unique new set of molecules. These give some cooked foods their flavors, scents, and colors.

Caramelization—turning sugar into caramel—is another reaction that only happens when heat is applied. Water molecules are driven off from sugars as steam, converting them into new forms of sugar and eventually into other molecules that produce a nutty flavor and brown color. Together, the Maillard reaction and caramelization produce the distinctive flavors, smells, and colors of cooked foods like freshly toasted bread, browned meat, roasted coffee, and popped corn.

So why do humans find these aromas, tastes, and sights so appetizing? It's because among the chemicals released during cooking are many scent molecules similar to those given off by ripening fruit.

Fruits are a source of energy-rich sugars that would have appealed to our ape-like ancestors as an essential part of their diet.

HOW COOKING MADE US HUMAN

Anthropologists claim that mastering the chemistry of cooking was the driving force behind human evolution. Cooking makes food easier and quicker to eat and digest, releasing more calories and allowing a wider range of foods to be added to the diet. Cooked food, especially meat, needs less chewing and uses less energy to digest, and it seems cooking allowed early humans to evolve energy-hungry large brains while freeing up time to invest in culture, society, and technology. Cooking explains why our ancestor species *Homo erectus* (reconstructed model pictured) evolved a smaller jaw, shorter intestines (and therefore smaller belly), and a larger skull (to accommodate that big brain).

Medicines, Mummies, and Makeup

The word "chemistry" traces its roots back to ancient Egypt, where the civilization that developed in the Nile Valley from 3100 BCE matched the sophistication of its art and architecture with equally sophisticated chemistry. Ancient Egyptians used a wide range of chemicals and knew how to refine and combine them to best effect.

Metals and Mysticism

The ancient Egyptian world was one of vivid colors, produced via a mastery of pigments and dyes used to enrich paintings, fabrics, makeup, and glass. The Egyptians took prehistoric pigments like ocher and other iron oxides, and added pigments based on cobalt, lead, and copper, using more elements than had been used previously. Lead, for instance, was mined in the form of galena, an ore found in Gebel Rasas ("the Mountain of Lead") a few miles from the Red Sea coast. Mercury, also known as quicksilver, was mined too. Egyptian sages began to develop a complex system of mysticism and lore relating to the known metals, linking gold to the Sun, iron to Mars, copper to Venus, and lead to Saturn. Although not scientific, this

Galena is the main ore of lead, which Egyptians used in glass, cosmetics, and medicines.

knowledge system followed its own logical rules, so it might be said to mark the beginnings of chemistry.

Mixing Pigments

Silicon is the second most abundant element in the Earth's crust after oxygen, and the Egyptians put it to good use. As early as the 16th century BCE they had developed furnaces hot enough to melt it and produce glass, and they later learned how to add lead to glass to make it sparkle.

A technology parallel to glass was the manufacture of faience, a ceramic made from a paste of

Blue faience statuettes of hippopotami were placed in the tombs of high-ranking Egyptian officials. Faience was made from silica, along with small amounts of calcium and sodium, and is known for its bright colors, especially blue, green, and turquoise.

crushed quartz (an oxide of silicon in crystal form) or sand. To this they added small amounts of lime (calcium carbonate) and natron, a naturally occurring mix of sodium carbonate, sodium bicarbonate, and common salt. Natron was a sort of secret ingredient in Egyptian chemistry; it helped lower the melting point of the quartz and made the glass more moldable. Faience was glazed with copper pigments to give a bright blue-green color, forming an artificial substitute for rare and expensive lapis lazuli.

The Egyptians also used natron and silicon to create an entirely new color, Egyptian blue. This artificial pigment was made by heating a mixture of sand, natron, and copper filings to about 1,500°F (850°C). And from the Levant, in the eastern Mediterranean, the Egyptians obtained royal purple, a purple dye made from a snail. Other pigments known to have been used in Egypt at least as early as 2650 BCE included charcoal, copper carbonate, and limestone.

ORIGINS OF THE WORD "CHEMISTRY"

The word "chemistry" comes from alchemy (see pp. 38-39), the Western pronunciation of the Arabic *al-kimya*, but the origins of the root word *kimya* vary according to the source. The Roman natural philosopher Pliny the Elder claimed it derived from the ancient Egyptian *kemi*, meaning "black"—the name first given to Egypt itself because of the fertile black silt of the Nile. Alternatively it was said to derive from the Greek word *khemeia*, meaning "pour together," referring to the art of melting metals.

Cure and Preserve

The pharmaceutical use of chemicals probably predates humanity, as there's evidence that apes heal themselves using medicinal plants. But the chemistry of medicine reached new heights under the Egyptians, who used many chemicals that stayed in the pharmacopoeia, an official list of medicinal drugs and preparations, until the 19th century. The Ebers Papyrus is one of the two oldest medical treatises and shows that the ancient Egyptians were familiar with medicines based on lead and antimony (a silvery-white metal used to treat fevers and skin conditions), as well as a host of plant extracts such as opium and aconite.

Written in about 1543 BCE, but based on much older sources, the **Ebers Papyrus** contains prescriptions written in hieroglyphics for over 700 remedies, including eye ointments containing lead and antimony.

When remedies failed, the Egyptians were experts in the chemistry of the mortuary arts. Their recipe for mummification made use of the properties of natron (a mix of sodium carbonates), which absorbs water to dehydrate the corpse and has antibiotic properties. Sodium carbonates are alkalis (see pp. 60–61), which raise the pH (alkalinity) of the treated flesh and help delay bacterial growth. The dried corpses were then treated with pitch and tars, such as bitumen, to seal and preserve them still further. In the right conditions, mummies can survive intact for at least 3,000 years.

Anubis, the ancient Egyptian god of the dead, was believed to oversee **embalming and mummification**. The entire process took about 70 days.

A 3,300-year-old painted stucco-coated limestone bust of Nefertiti (ca. 1370–ca. 1330 BCE). The Egyptian queen is said to have used **lead-based cosmetics** to enhance her renowned beauty.

The electron configuration of **lead** (Pb).

Pb

TOXIC COSMETICS

It's known that Queen Nefertiti and other ancient Egyptian royals used lead-based substances to make cosmetics, especially as an ingredient in black eye makeup (kohl). But nowadays lead is banned in cosmetics on the grounds of its high toxicity.

So how much lead was in ancient Egyptian cosmetics? By analyzing cosmetics found in ancient tombs and recreating ancient recipes, chemists have shown that two non-naturally occurring lead chlorides, laurionite

and phosgenite, must have been made by the ancient Egyptians and used as fine powders in makeup and eye lotions. According to ancient manuscripts, these were essential remedies for treating eye and skin ailments. However, laurionite contains almost 80% lead and phosgenite is 76% lead!

Matter and Energy

Let's take a brief time out from the ancient world and look at some basic concepts and terminology that will help you to understand the following pages. Chemistry is the study of matter—its makeup and transformations—so its key concepts concern the description of matter. Since chemical transformations involve energy, ideas relating to energy form the other key concepts of chemistry.

Phases of Matter

Matter is anything in the universe that has mass and takes up space; it's the material of the physical world, the stuff you can see and touch. Matter can exist in three states, known as phases: **solid**, **liquid**, and **gas**:

Solid: A solid has a definite shape and volume, because the particles that make it up, whether atoms or molecules, are held close together in a fairly rigid structure by powerful bonds (which may be covalent or ionic—see pp. 64–65). Sometimes this structure has a repeating pattern

called a crystal lattice. Solids with a crystal lattice include ice, table salt, granulated sugar, and quartz. The particles in a solid aren't completely motionless; they vibrate in place, but remain fixed relative to one another.

Liquid: When a solid is heated to its melting point, it becomes a liquid. In this phase, matter has no definite shape but still has a definite volume—in other words, the liquid remains as a definite body or mass. In a liquid the bonds, or forces of attraction, between particles are stronger than in a gas but much weaker than in a solid, allowing the particles to move about.

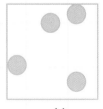

melting ⇌ freezing

evaporating or boiling ⇌ condensing

solid (s)   liquid (l)   gas (g)

Gas: When a liquid is heated to its boiling point, it becomes a gas, the phase in which matter has no definite form or volume. In a gas the forces attracting particles together are so weak that the particles can move about freely. As a result a gas will expand to fill the volume available.

"Melting" and "boiling" describe phase changes from solid to liquid and liquid to gas. The reverse phase changes are known as "freezing" (from a liquid to a solid) and "condensing" (from a gas to a liquid). Some substances go directly from solid phase to gas phase, which is called "sublimation." Solid (frozen) carbon dioxide, known as dry ice, is an example of a substance that sublimates. However, the misty white vapor given off by the dry ice isn't carbon dioxide gas, which is colorless and odorless, but water vapor condensing out of the air as it is cooled by the subliming carbon dioxide.

The particles in **ice (the solid form of water)** are held together in a repeating pattern known as a crystal lattice.

Phase changes are caused by the amount of energy the particles have. High levels of energy enable particles to break free of bonds that hold them together in their solid phase and change to liquid or gaseous phases. As a substance is cooled, the particles lose energy and the attractive forces between the particles reassert themselves.

Frozen carbon dioxide is more commonly known as dry ice. The smoky vapor given off is often mistaken for gas, but it is actually water vapor condensing out of the air.

Properties of Matter

••••••••••••••••••••••••••

Matter can be pure or mixed. In a mixture, different substances are physically combined—for instance, if you mix together chalk dust and salt, or dissolve salt in water. A pure substance has a single, constant composition, and can be either an element or a compound (see p. 9). Pure substances have chemical and physical properties, and these are what chemists study:

Chemical properties include how reactive a substance is, what it will react with, and other factors affecting the substance's transformation into other substances.

Physical properties include mass, dimensions, volume, density, color, conductivity, and so on. These are described with standard measurement units, such as the gram, meter, and liter. Smaller units have prefixes, such as "centi-" ("a hundredth of") and "milli-" ("a thousandth of")—a milligram is therefore 0.001 grams. Volume is measured in cubic meters (m^3), cubic centimeters (cm^3), cubic millimeters (mm^3), or liters (l)—more commonly milliliters (ml). Density is mass divided by volume, so is typically measured in g/ml.

When **copper sulfate** meets fire, an interesting reaction occurs. The heat from the flame excites the electrons enough that they give off energy—seen as a green photon of light.

ESSENTIAL ENERGY

Along with matter, energy is one of the two basic components of the universe. Energy can take different forms, and the most important forms in chemistry are kinetic and potential energy:

Kinetic energy is the energy of motion that particles have, and determines the speed and force of their motion, governing properties such as phase and reactivity.

Potential energy is the energy stored in a substance, which can become other types of energy. In chemistry the energy stored in the form of chemical bonds is of most interest. Breaking the bonds takes energy but can also release energy. Except in nuclear reactions, energy can't be created or destroyed, only converted from one form into another.

Potential energy can be transformed into **kinetic energy**. A molecule such as methane has potential energy stored in bonds between the atoms. When the methane is ignited, the bonds of the atoms break, and this releases kinetic energy such as heat and light.

Natural Philosophy in Ancient Greece

Around the 5th century BCE, new ways of thinking about the natural world emerged in ancient Greece. Although older civilizations such as the Egyptians and Babylonians had extensive practical experience of chemistry from medicine to metallurgy, they had made no attempt to study and explain the natural world. This was about to change with the development of the first theories concerning the makeup of matter.

Evolving Elements

The ancient world made use of gold, silver, tin, lead, copper, iron, mercury, antimony, sodium, calcium, carbon, sulfur, and arsenic in one form or another. Yet they weren't recognized as elements, and not until the Greeks was there any attempt to analyze and explain the differences between these and other forms of matter.

Among the first to use evidence from the natural world to ask and answer fundamental questions about the basic features of matter was the natural philosopher Thales of Miletus (ca. 625–ca. 547 BCE), a semi-legendary figure from a Greek city-state in what is now Turkey. Thales is credited as being the first man to seek natural explanations instead of attributing phenomena to the gods. He also defined general principles, identifying water as the first principle—the basic element from which all other matter is composed.

Thales of Miletus, one of the Seven Sages of Greece, is considered by many to be the first "natural philosopher."

Thales was followed by his pupil Anaximenes, whose career flourished ca. 546–526 BCE. He argued that the fundamental element was in fact air, and that through condensation and rarefaction (an increase and decrease in the density of air molecules), it became earth, fire, water, and all the other forms of matter. Reacting against this theory, Heraclitus of Ephesus (ca. 500–475 BCE) proposed that nature was in a state of constant change and therefore the basic element must be changeable, so he identified fire as the first principle.

Finally, Empedocles (ca. 492–432 BCE) claimed that there were four basic forms of matter—earth, air, fire, and water. Thanks to its adoption by Aristotle (see pp. 30–31), this "four element" model would become the accepted standard for over 2,000 years. It can be argued that the four elements correspond to the modern model of the universe, with three phases of matter (earth = solid, water = liquid, air = gas), together with energy (= fire).

Thanks to its adoption by Aristotle, the "four element" model would become the accepted standard for over 2,000 years.

STEAM POWER

Hero of Alexandria (62-152 CE) was a Greek natural philosopher and inventor who was millennia ahead of his time in pneumatic chemistry—the study of gases (see pp. 56-57). He is most famous for his aeolipile (see below), a device he built using his knowledge of the phase change between water and steam. A cauldron of boiling water supplied steam to a sphere with two vents, and as the steam was forced out under pressure, the sphere rotated. It was the first steam engine, a technology based on pneumatic chemistry that would later reshape the world by triggering the Industrial Revolution. So why didn't Hero's invention trigger a revolution in his own time? Probably because slaves were widely used, so there was no need for labor-saving devices at this point in history.

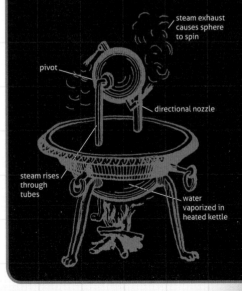

steam exhaust causes sphere to spin

pivot

directional nozzle

steam rises through tubes

water vaporized in heated kettle

The Birth of Atomic Theory

The foundation of modern chemistry is atomic theory—the model that explains what substances are made of, and how and why they come together to form molecules. Atomic theory in its modern form dates to the 19th century, yet it drew on an ancient but largely forgotten tradition—the atomists of ancient Greece.

Invisible Particles

A school of Greek philosophers known as the Eleatics argued that since nothingness is logically impossible, there can't be any space between particles and therefore no separate, indivisible particles. This complex argument led to apparently irrational conclusions—for instance, they claimed that change is impossible. Reacting against this school of thought, the philosopher Leucippus (5th century BCE) and his pupil Democritus (ca. 460–370 BCE) claimed that empty space—today known as a vacuum—can exist, and therefore it's possible for particles to exist.

These particles are unchangeable and indivisible, known as atoms (from the Greek *atomos*, meaning "uncuttable"). In other words, if you cut a piece of matter into smaller pieces, and then cut up those pieces, and so on, you will eventually reach the smallest unit possible—a particle

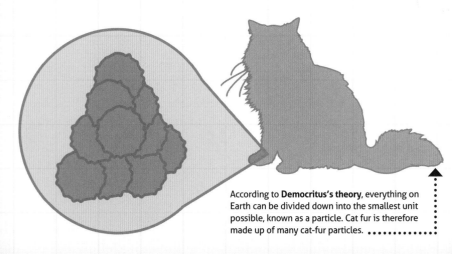

According to **Democritus's theory**, everything on Earth can be divided down into the smallest unit possible, known as a particle. Cat fur is therefore made up of many cat-fur particles.

that can't be cut into smaller particles. According to this "atomic theory," atoms are solid and so small that they're invisible, but they come in different shapes and sizes and can change position. According to Democritus, different arrangements and combinations of atoms produce different materials, even different worlds.

Rejected Theory

The atomists, as Leucippus, Democritus, and their later supporters were known, seem to have been astonishingly prophetic, anticipating modern thinking about atoms, elements, and cosmology (the study of the universe). However, their model was just speculation, and not based on a true scientific method (see pp. 66–67). It included mystical, or metaphysical, ideas, such as the belief that the human soul was also made of fine, round atoms.

Although atomism had its supporters in ancient Greece, it was rejected by the most influential of later philosophers, such as Plato and Aristotle, and didn't become popular again until the scientific revolution of the 17th and 18th centuries. Whether Democritus' insights into the nature of matter would have speeded the arrival and advance of scientific chemistry is impossible to say. But the theories that took the place of atomism, especially Aristotle's ideas, have been held responsible for leading chemistry up a 2,000-year-long blind alley.

LIVING, OR DYING, BY PRINCIPLES

Although Democritus' theories failed to attract support, he was fortunate to live to a great age. Maybe this was thanks to his personal philosophy, which was that cheerfulness should be the purpose of life; to later generations he was known as "the laughing philosopher." Other ancient Greek philosophers mentioned in this chapter weren't so lucky. Thales of Miletus was said to have fallen off a hill because he was so intent on studying the stars, while Heraclitus met perhaps the most colorful end.... Because of his philosophical principles, he kept to a starvation diet, which caused him to swell up with dropsy (edema). Seeking to expel his "ill humors," he buried himself in a dung heap and never came out.

Democritus, advocate of **the atomic theory**, known to later generations as "the laughing philosopher."

Introducing Atoms

At the scale of matter investigated by chemistry, the basic building block is the atom. Atomic structure determines the properties and chemistry of a substance, and this structure is made up of subatomic particles. It's now known that Democritus and the ancient atomists were wrong, and that far from being indivisible, atoms actually have complex internal workings.

Atoms and Elements

An atom is the smallest particle of an element. All the atoms of an element are identical to each other (except isotopes—see pp. 132–133) and different from the atoms of other elements. Each element has a unique atomic structure, and this defines the element and influences its nature.

Subatomic Particles

Atoms contain three subatomic particles: protons, neutrons, and electrons. Protons and neutrons are far larger than electrons, and together make up over 99.99% of an atom's mass. The number of protons determines the atomic number, while the combined total of protons and neutrons determines the mass number (see pp. 98–99 for more on atomic number and mass).

These subatomic particles may have an electrical charge: protons are positive (+1), electrons are negative (−1), and neutrons have 0 charge. An atom is electrically neutral, because it has an equal number of protons and electrons. If an atom loses or gains an electron, it may become positively or negatively charged, and is known as an ion.

The Orbital Model

The simplest model of an atom's internal structure is the orbital model, in which the atom is like a tiny solar system. At the center is the

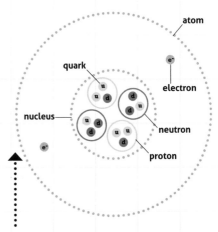

The orbital atomic model, developed by Danish scientist Niels Bohr, in which electrons orbit the nucleus. The nucleus is made up of protons and neutrons, which are themselves made of quarks.

nucleus, where the protons and neutrons are packed together. Circling around the nucleus are the electrons, arranged in orbits, or shells, at different distances. These shells represent different energy levels, with the shell nearest the nucleus having the lowest energy level. Electrons can move between shells, but there's only so much space in each shell. This arrangement determines the atom's valency (combining power), which in turn determines its chemical properties (see pp. 64–65).

In real life the picture's a little more complex. Closer to what scientists see is the quantum mechanical model, in which it's impossible to know both the position and momentum of an electron at the same time. Consequently electrons are said to occupy fields of space known as orbitals or electron clouds.

Two's Company
....

For most of the history of chemistry it wasn't possible to deconstruct every element to single atoms. Some elements, even in their pure state, favor binding to another atom. For instance, oxygen never exists as

A SUBATOMIC ZOO

There are so many subatomic particles, they are collectively known as the "particle zoo," though apart from protons, neutrons, and electrons, they don't really affect the chemistry of matter. The zoo includes many strange species such as antiparticles—mirror images of particles. For example, the antiparticle of an electron is a positron. Over 200 subatomic particles have been discovered so far, divided into quarks, leptons (including the electron), and bosons.

single atoms of oxygen—even in a jar filled with nothing but pure oxygen, the element will exist as pairs of atoms, bound together in a "diatomic" molecule. Six other elements behave in a similar way: hydrogen, nitrogen, fluorine, chlorine, bromine, and iodine. This phenomenon caused many headaches for 18th- and 19th-century chemists trying to work out atomic numbers and masses.

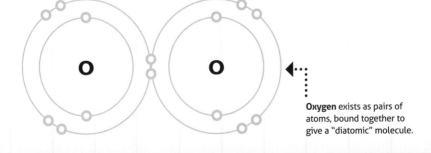

Oxygen exists as pairs of atoms, bound together to give a "diatomic" molecule.

Aristotle

Towering over the history of science from ancient times to the early modern period is the colossus Aristotle, whose model of the physical world and theory of the elements were accepted as standard until the 16th century and beyond. The student of Plato and tutor to Alexander the Great, Aristotle became a legendary figure, yet for chemistry his legacy was double-edged.

Students and Masters

Aristotle (384–322 B CE) was born in Macedonia, northern Greece. At age 17 he traveled to Athens to study with the philosopher Plato at the Academy, and soon proved himself to be a brilliant student. Aristotle remained at the Academy until Plato's death, then took up positions in Asia Minor (part of present-day Turkey) and the Greek island Lesbos, during which time he pursued research into marine biology that was unmatched until the modern era. In 342 B CE he was headhunted by Philip II of Macedon to act as tutor to the young prince, Alexander.

When Alexander came to the throne in 335, Aristotle returned to Athens to found a school of his own, the Lyceum. But anti-Macedonian feeling ran high in the city and when Alexander died in 323, Aristotle decided to flee to the city of Chalcis, where he died the next year.

Logical Assumptions

Plato had disliked the realities of the material world, and the practice of observing it rather than simply thinking about it, and one of Aristotle's most important innovations was to get his hands dirty in research. Yet the heart of Aristotle's philosophical project was his system of logic, where knowledge was gained by logic and rational thought alone. He formalized a system of deductive logic that used syllogisms. In a syllogism you start with two premises (propositions) and use them to deduce a conclusion. But if the premises are faulty, then the conclusion will be faulty. This may explain how Aristotle came to believe that the universe was made up of five elements.

Accepting the four terrestrial elements of Empedocles (see p. 25), he added a fifth one—ether—to explain the workings of the heavens. For Aristotle the natural qualities of the elements explained

One of Aristotle's most important innovations was his willingness to get his hands dirty in some fields of research.

everything about the material world. Earth was naturally heavier than air, so substances with a higher proportion of the earth element would naturally fall until they were lower than airy substances. Substances with a high proportion of the fire or water element would have "hot" or "damp" qualities respectively, and this explained their chemistry. But Aristotle had mistaken qualities for properties (in the scientific sense), and this fundamental error made his conclusions incorrect.

LONG-LASTING LEGACY

Aristotle's theories about the elements were so important because of the huge influence his work would have on natural philosophy, particularly in Europe, over the next 1,900 years. For a variety of reasons his system of logic and physics appealed to the Church, and this in turn preserved his popularity and authority. By the Middle Ages (ca. 1100-1453 CE), Aristotle was the focus of scholasticism, a system of thinking that dominated the intellectual life of Europe, and he was taken as the last word on natural philosophy, including chemical matters.

He made similar errors in other fields, after failing to back up logic with observation. For instance, relying on deduction, he concluded that the function of the brain is to cool the blood and that man has only eight ribs on each side. Today it seems incredible that Aristotle could have made simple mistakes, such as his claim that women have fewer teeth than men.

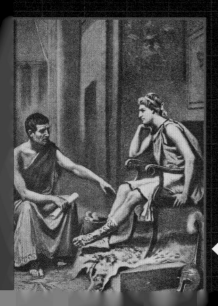

Aristotle was tutor to Alexander the Great

Chemical Warfare: Greek Fire

A remarkable example of ancient chemical technology was the secret weapon that preserved the Byzantine Empire for 600 years—the mysterious "Greek fire." The recipe for this extraordinary chemical weapon was among the most closely guarded secrets in history, and with good reason, as it had the power to change the very course of history.

Smells Like Victory

Greek fire was a napalm-like incendiary substance used in defense of the Byzantine Empire, the Greek-speaking eastern part of the Roman Empire. The technology of Greek fire was a closely guarded secret, known only to the imperial family and associates, and it remains an enigma to this day. It was first used in 678 CE against the Arabs, who were threatening to overrun Constantinople, having already conquered the Persians. Although the defenses of the capital could hold back land armies, the city could be forced to surrender if the Arab fleet gained control of the seas.

But the Arabs had sewn the seeds of their own defeat. When their armies overran Christian Syria, refugees flocked to the safety of Constantinople. Among them was a Syrian Greek called Kallinikos, who brought with him the recipe for a secret weapon that would become known as "Greek fire"—"liquid fire" and "sea fire" were alternative names. Incendiary weapons based on petroleum products, such as pitch or naphtha (a flammable oil), were part of the Arab arsenal. In fact they were probably known in one form or another to the Romans and Persians

Found in an illuminated manuscript, this is one of the few surviving ancient depictions of **Greek fire**.

before them. What distinguished the new "Greek" fire was its advanced composition and, crucially, the apparatus used to spray the flaming liquid toward the enemy.

Liquid Death

Even today it's only possible to speculate on the composition of Greek fire, but it's generally thought to have included sulfur, quicklime (calcium oxide), liquid petroleum, and perhaps even magnesium (part of modern incendiary weapons). Magnesium is a highly reactive metal that will even burn underwater, one of the characteristics of Greek fire, which helped to make it such a fearsome weapon. To spray this liquid death, the Byzantines invented an ingenious siphon device.

The effects of Greek fire were devastating. In 678 CE it shattered the Arabs' navy, killing thousands of men. The blockade was broken and the Arabs were forced to appeal for peace. When they attacked again, in 717, Greek fire once more played a key role, and the Arabs were again beaten back with severe losses.

Over the next 300 years, Greek fire was vital in defense of the Byzantine Empire, but by 1204 the secret had somehow been lost. Incendiary weapons were still used, but the technology that made Greek fire so formidable was no longer available. The empire fought on for five more centuries, until the Ottoman Turks broke through Constantinople's walls with gunpowder in 1453.

The technology of Greek fire was a closely guarded secret, known only to the imperial family and associates, and it remains an enigma to this day.

A HOLY SECRET

Greek fire was used as little as possible, to help prevent the siphon apparatus falling into enemy hands. Writing to his son, Emperor Constantine VII Porphyrogennetos stressed that the secret must not be revealed even to allies, explaining: "The ingredients were disclosed by an angel to the first great Christian emperor, Constantine … [who] ordained that they should curse, in writing and on the Holy Altar of the Church of God, any who should dare to give this fire to another nation…."

The shape of Europe and the direction of world history might have turned out very differently without the secret of Greek fire.

2

Alchemy
and the Birth of
Chemical
Science

The chemical legacy of the ancient world was a colorful stew of materials and mysticism that proved irresistible to scholars through the ages. It drew the greatest minds of late antiquity, medieval Islam and the European Renaissance (14th–17th centuries) like moths to a flame. Their efforts to unlock the secrets of the cosmos would lead to a new way of investigating nature and toward a new science of matter and its transformations.

The Roots of Alchemy

In Egypt in 331 BCE, Alexander the Great founded the city of Alexandria. It came to represent his new world, a thriving, cosmopolitan blend of races, cultures, and traditions. It was in Alexandria that chemistry would begin to emerge in the form of alchemy—an art rather than a science, but recognizable in many of its practices and concerns. Like the city itself, this new art was a complex hybrid.

Mystical Wisdom

In Alexandria under the Ptolemaic dynasty (305–30 BCE), Greek culture was combined with ancient Egyptian traditions. Magic, mysticism, and the chemical arts that the Egyptians had practiced for millennia, such as embalming, glass-making, and metallurgy, went into the crucible with Greek metaphysics and Aristotle's cosmology.

What emerged was a strange new compound. Alchemy adopted the classical elements of earth, air, fire, water, and ether, and tried to make use of Aristotle's theories about matter. If the nature of a substance was controlled by its proportions of elements, as Aristotle said, then altering those proportions would change the nature of the substance. If gold, for instance, was formed by one recipe of earth, air, fire, and water, then taking a baser metal such as lead and altering its recipe until it matched that of gold, should transmute lead into gold. Seeking to perform this transmutation,

HERMES TRISMEGISTUS

The father of alchemy was said to be the semi-divine Hermes Trismegistus ("thrice greatest" Hermes), a fusion of the Egyptian god of wisdom, Thoth, and the Greek god Hermes. He was said to have written a work known as the *Hermetic Corpus*, and all subsequent alchemical research was an attempt to decipher and recover this original wisdom

alchemists worked with substances known from Egyptian, Greek, and Roman technology: the metals, earths (including ocher), and ores, performing operations such as solution, distillation, and filtration.

Guiding these operations were mystical laws, especially "as above, so below." This saying represented a belief that the microcosm—the "little" world of man and earthly matter—reflects the macrocosm, or the universe, including the stars and heavenly bodies. This law gave rise to the belief that everything in the microcosm, or terrestrial sphere, mirrors everything in the macrocosm, or heavenly sphere. The seven known metallic elements, for instance, were associated with heavenly bodies—with gold linked to the Sun, silver to the Moon, copper to Venus, and so on. Equally, plants, gemstones, star signs, and all other natural and human phenomena fitted into a network of connections, and these could be used to influence the substances that alchemists worked with.

CHINESE ALCHEMY

China had a tradition of alchemy at least as ancient as the Western one. Chinese alchemists were especially interested in prolonging life, and a number of Taoist sages were said to have discovered elixirs of immortality. As well as "internal alchemy" dedicated to human health, the Chinese practiced "external alchemy" with similar aims to the Western version—the production of gold. A probable side-product of research was gunpowder, and according to a Taoist alchemical text of ca. 850 CE, it was an explosive discovery: "Some have heated together the saltpeter, sulfur, and carbon of charcoal with honey; smoke and flames result, so that their hands and faces have been burnt, and even the whole house burnt down."

Unscientific Alchemy

Such mystical wisdom was believed too powerful to be shared with the uninitiated, so alchemists recorded their knowledge using symbols and allegories—stories that revealed a hidden meaning. This secrecy was just one way in which alchemy was unscientific. Alchemy was also highly subjective and often mystical in nature—achieving a good result depended on variables such as the experimenter's state of spiritual purity, while substances were affected by the phase of the Moon or the position of the stars.

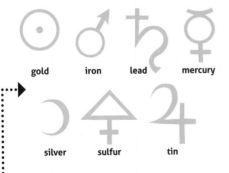

gold   iron   lead   mercury

silver   sulfur   tin

Alchemists used a **series of symbols** in order to record their findings. They were used to denote some elements and compounds up until the 18th century.

Chemical Reactions: The Basics

In their quest for the transmutation of substances, the alchemists were making a category error. They thought they were achieving what is today known as a nuclear transmutation—the transformation of one element into another element. But actually they were performing chemical reactions—the processes by which compounds (elements bonded together in various forms) are created, destroyed, or changed.

What Happens?

In a chemical reaction, a substance or mix of substances is changed into different substances. The substances at the start of the reaction are known as **reactants**, and the substances left at the end are called **products**. Chemical equations are used to show reactants and products, with an arrow showing the direction of the reaction, like this:

reactant A + reactant B → product AB

Rusting is an example of a chemical reaction, where iron combines with oxygen to form iron oxide (rust). The reaction can be written like this:

iron + oxygen → iron oxide

Here's another example, showing what happens when you light the gas on your stove:

methane (g) + oxygen (g) → carbon dioxide (g)

The phase of the substance (gas, liquid, or solid) is shown in brackets. Here, all the reactants and products are gases.

Here's an example of a reaction that an alchemist would have performed —the reduction of silver:

$$silver\ (l)\ +$$
$$silver\ carbonate\ (s) \rightarrow carbon\ dioxide\ (g)$$
$$\uparrow + oxygen\ (g)$$
$$heat$$

And here's the reaction that gave Chinese alchemists such a nasty shock when they accidentally made gunpowder in the 9th century (see p. 37):

sulfur (s) +   carbon dioxide (g) +
carbon (s) + → pot. sulfide (s) +
pot. nitrate (s)   nitrogen (g)

Instead of writing out names in this way, science adopted a universal language (see pp. 94–95) based on Greek and scientific notation to improve accuracy and save time.

Reactions that generate heat are known as **exothermic**, while those that absorb energy are known as **endothermic**. In the examples here, the combustion of methane and gunpowder are exothermic, while the reduction of silver is endothermic. Some reactions happen spontaneously (for example, metals rusts in air or water), but many reactions, including exothermic ones, need an initial input of energy known as "activation energy." Once this activation energy is supplied, an exothermic reaction generates enough energy to keep itself going.

Balancing Out

An important principle of chemistry is that matter cannot be created or destroyed (except in nuclear reactions). This is known as the "law of conservation of matter." In reaction equations, this means there must be the same number of atoms on either side. When writing an equation it's necessary to make sure it balances, and this is where chemical notation comes in so handy (see pp. 110–111).

REACTION TYPES

Combination reaction: where two or more reactants combine to produce a single product.

Decomposition reaction: where a single reactant breaks down to give two or more products (the opposite of a combination reaction).

Displacement reaction: where a more active element takes the place of a less active one from a compound. Metals in particular have a hierarchy of reactivity: alkali metals like sodium and magnesium are at the top, followed by aluminum and zinc, with copper, silver, and gold at the bottom being the least reactive. So if you add zinc to a solution of silver salt dissolved in water, the zinc will displace the silver, which will precipitate out as metallic silver. But if you then add aluminum to the solution, the zinc will be displaced and precipitate out.

Combustion reaction: where a compound combines with oxygen—otherwise known as burning. Combustion is a common example of a class of reactions called "redox"—short for reduction-oxidation. In a redox reaction, electrons are swapped between reactants (see pp. 90-91); rusting (pictured far left) is another example of a redox reaction.

Chemistry in Medieval Islam

The next stages in the evolution of chemistry took place in the Middle East, in the world of medieval Islam. During a golden age of intellectual activity, a series of great names in alchemy created new principles, perfected techniques and processes, and preserved and built up a great store of knowledge that would profoundly influence the European Renaissance.

Center of Learning

Persia (modern-day Iran) had long been a center of scholarship where ideas and influences from East and West met and merged. This was thanks to the Silk Road, an ancient network of trade routes that connected China to the Mediterranean Sea. In late antiquity the region received an influx of classical knowledge from Christian scholars driven out of the Byzantine Empire, who established theological and medical schools.

The arrival of Islamic rule in the 7th century brought rapid and dramatic change to the region. Under the Abbasid Caliphate of the 8th–11th centuries, learning flowered in the Islamic world. Ancient Greek literature was translated into Arabic, alongside important Indian and Chinese works. Scholars flocked from across the Islamic Empire and beyond to the Abbasid capital at Baghdad, where institutions such as the House of Wisdom encouraged the study of mathematics, astronomy, medicine, chemistry, zoology, geography, alchemy, and astrology. The rapid growth of scholarship was helped by high levels of literacy and the rise of the use of paper.

▲ A 13th-century illustration of scholars learning at an **Abbasid library**.

Building on Ideas

The collapse of the Roman Empire in Europe meant that most ancient books were lost. But the Islamic world kept classical learning alive and developed it further. Islamic alchemists drew on classical sources such as Aristotle and his elements, the physician Galen and his "four humors" theory—the belief that health depended on a balance of four bodily fluids (humors)—and Chinese and Indian alchemists' work.

The first great name was Jabir ibn Hayyan (ca. 721–ca. 815 CE), the "father of Islamic alchemy" (see pp. 42–43). He was succeeded by the Persian physician Al-Razi (ca. 865–ca. 925 CE), whose writings are considered revolutionary because he began to approach a scientific philosophy. He was more willing to study substances in isolation, relying on his observations of what actually happened, rather than on theories, and his book, *The Secret of Secrets*, would become a bible for European alchemists (see box).

After Al-Razi came Abu Ali ibn Sina (980–1037 CE), who specialized in alchemical medicine. He developed Galen's four humors theory and became an important influence on the first wave of early Renaissance scholars in Europe.

THE SECRET OF SECRETS

The Secret of Secrets was set out almost like a manual, with a section describing all the exotic glassware that had been perfected by Islamic alchemists, and which would remain standard equipment in chemistry laboratories until the 19th century. In the last section of the book, Al-Razi attempted to classify substances, perhaps starting a process that would end in the periodic table. Jabir decided that all metals are formed from sulfur and mercury, and Al-Razi added a third constituent—salt. This theory would greatly influence Paracelsus (see pp. 52-53).

A page from *The Secret of Secrets*, with charts for determining whether or not a person will die.

Jabir ibn Hayyan

Jabir ibn Hayyan was the first great figure in Islamic alchemy, though his achievements were not restricted to this field. His bold adaptation of classical models of chemistry greatly influenced later generations of alchemists. But perhaps more important were the advances he made on the practical and experimental side, discovering new substances, mastering new techniques, and developing chemical knowledge.

He argued that four "natures" (qualities) underlay Aristotle's elements—hotness, coldness, dryness, and wetness; and pairs of these qualities created the four earthly elements (for example, hot + dry = fire).

He particularly focused on the nature of metals, which he believed were formed from the elements sulfur and mercury. The proportions of each determined the nature of the metal, and the perfect balance produced gold. Jabir was convinced he could find a way to transmute lead into gold—by separating lead into sulfur and mercury, cleansing them of impurities, then reconstructing them with new proportions. To achieve this, he would need to use a substance that played a part in the process but remained unchanged itself—today known as a catalyst.

Natures and Metals

Jabir came to alchemy from the practical side, having practiced as an apothecary—the medieval equivalent of a dispensing chemist. However, he took a holistic view of learning, seeing alchemy as just one aspect of natural philosophy.

Jabir drew inspiration from the alchemical book the *Emerald Tablet*, said to be written by Hermes Trismegistus (see p. 36), and from Aristotle's elemental theory, and then added new dimensions.

Practical Solutions

It was in applied chemistry that Jabir made the greatest

ENSNARED IN POLITICS

Jabir led an eventful life in an eventful era, as he practiced during the reign of the caliph Harun al-Rashid, the legendary ruler of *Arabian Nights* fame. Born in Persia (modern-day Iran) but of Arabian descent, Jabir was immersed in the dangerous world of caliphate power politics from the start. His father was executed for plotting to overthrow the Umayyad caliphate, the first Islamic dynasty. Jabir himself was closely associated with Harun's vizier, Jafar, so his fortunes rose and fell with his patron. When Jafar lost favor and was executed, Jabir was forced to flee Baghdad and retire to the country, where he spent his last days penning his massive book *The Sum of Perfection*, one of hundreds of books attributed to him.

Harun al-Rashid receiving a delegation sent by Charlemagne at his court.

contribution. He improved glass-making, the refining of metals, and the fabrication of dyes and inks—for instance, he developed an ink using iron pyrite (fool's gold, pictured left) for illuminating manuscripts.

He also invented a new acid—*aqua regia* (royal water, pictured right). Now known to be a combination of hydrochloric and nitric acids, this new acid had the power to dissolve gold, which can't be achieved with single acids alone.

Jabir introduced organic substances by experimenting on plant extracts

though at the time philosophers saw no clear divide between the organic and inorganic worlds. In synthesizing new compounds, Jabir was seeking to discover and even create new species. Perhaps most valuably, he recorded his experiments in a methodical way, describing materials, equipment, techniques, and results. Not only did this mark a proto-scientific approach, it meant his work could be a resource to later generations of alchemists.

Catalysts and Kinetics

Jabir's work with catalysts opened an important new chapter in chemistry. To understand it, we first need to look at kinetics—the study of reaction speeds. Kinetics examines the rate at which a reaction proceeds, and the factors that can affect this rate. Catalysts are one such factor. Another is temperature, affecting the speed of the reactive particles and therefore the rate at which they collide.

In **collision theory**, a chemical reaction happens when particles collide with enough force. In (a) the collision is weak so the reactant particles collide but bounce apart, unchanged. In (b) the collision is strong enough for a reaction to occur. **..................▶**

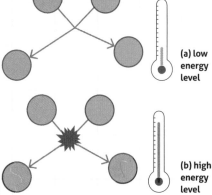

(a) low energy level

(b) high energy level

Collision Theory

The simplest model of how reactions occur is known as collision theory. According to this model, the atoms and/or molecules making up the reactants are like pool balls zooming around a table. The moving particles have kinetic energy. For a reaction to occur, they need to be moving fast enough (with enough kinetic energy) so that they collide with sufficient force to break chemical bonds and transfer their energy into new chemical bonds. This minimum energy that molecules need for a collision to result in a chemical reaction is known as the activation energy. As well as having sufficient energy, one reactive particle may also need to hit the other at just the right spot—at the particle's reactive site, where the collision must occur.

Turning Up the Heat

The temperature of a substance or mixture is a measure of the average kinetic energy of the particles making it up. Adding heat to a substance increases its temperature, because the heat energy is converted into kinetic energy and the average kinetic energy of the particles goes up. In other words, a higher temperature means that particles have more energy. Particles that have more energy are moving around more, and are therefore more likely to collide.

Another way of boosting reaction rates is to increase the concentration of the reactants. Having more reactive particles in a given volume of space increases the chances that they will collide, and the more collisions there are, the more likely a reaction will occur.

Little Helpers

. . . .

A catalyst is a substance that speeds up the rate of a chemical reaction but remains unaltered at the end of the reaction. Only a minuscule amount may be needed to produce a big effect. It's important to note that a catalyst does not increase the amount of end products or alter the balance of a reaction. There are two types of catalyst:

Heterogeneous catalyst: one that is in a different phase from the reactants—typically it would be a solid, while the reactants would be gases or liquids. This catalyst works by capturing one of the reactants and holding it in such a way as to present its reactive site, increasing the likelihood of another reactive particle colliding at the right spot and successfully reacting.

Homogeneous catalyst: one that is in the same phase as the reactants. These catalysts often work by offering an alternative reaction pathway for the reaction—one with a lower activation energy and faster kinetics. Typically the catalyst forms intermediate compounds at transition states in the new reaction pathway, before decoupling from the reactant and returning to its normal state.

For instance, if AB is the reactant, A and B are products, and C is the catalyst, the reaction pathway might look like this:

$$C + AB \rightarrow CAB \rightarrow CA + B \rightarrow C + A + B$$

This is comparable to the much simpler equation: $C + AB \rightarrow C + A + B$

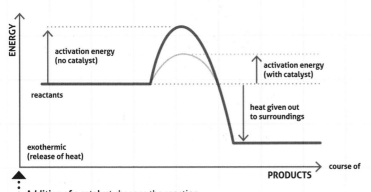

Addition of a catalyst changes the reaction pathway by forming intermediate products at lower energy levels, reducing the activation energy necessary to trigger the reaction.

Poisons and Poisoners

One of Jabir's many works was the *Kitab al-sumum* (*Book of Poisons*), a key work in toxicology (the study of poisons). Jabir and other Islamic proto-chemists were instrumental in passing on to medieval Europe the vast accumulated body of lore and practice concerning the use and treatment of poisons, one of the darker aspects of chemistry.

Deadly Arsenic

Our relationship with chemicals has always been a two-way street—we have long made use of chemicals, but we've been affected by them too. When a chemical is likely to harm an organism, it's said to be a toxin (from *toxicon,* the Greek for the poison used on arrow tips). Chemicals from plant, animal, and mineral sources have been known and used as poisons for millennia. The earliest known case of poisoning is probably that of Ötzi the Iceman, whose frozen body was found in an Italian glacier 5,000 years after his death. Analysis of his hair showed that he'd suffered from chronic arsenic poisoning, probably as a result of smelting copper ores contaminated with the toxin.

Arsenic wasn't identified as an element until 1250, by the alchemist Albertus Magnus (ca. 1200–1280), but its ores had been known for thousands of years before this. The ancient Romans, Indians, and Chinese all understood the potency of arsenic compounds as poisons and medicines. The Romans knew how to make the highly toxic salt sodium arsenite, and in fact there was an epidemic of arsenic murders in ancient Rome.

In this 1893 painting by John Collier, the **notorious poisoner Cesare Borgia** offers a guest some wine. Apparently Cesare died after mistakenly drinking his own poisoned wine.

But the golden age of poisoning, sometimes known as the Age of Arsenic, was Renaissance Italy, when the alchemists' expertise with arsenic compounds was widely misused. The most notorious poisoners of the late-15th century were the Borgias, infamous for inviting enemies to dinner and feeding them arsenic-laced dishes. However, Pope Alexander and his son Cesare Borgia were supposedly victims of their own game, accidentally drinking their own poisoned wine.

Murder for Sale

Over the next two centuries, poisoning became wildly fashionable. In 17th-century Rome a group of young wives were supplied with an arsenic-based brew known as *Aquetta di Perugia*, which they obtained from Hieronyma Spara, a reputed witch who was eventually executed. Soon after, another mistress of poisons, Giulia Tofana of Naples, started selling *Aqua Tofana*, just four drops of which would dispose of inconvenient relatives. Such poisons were known in France as *poudre de succession* ("inheritance powder"), because they allowed poisoners to come into their inheritances early.

The concoction and supply of toxins gave early chemists a bad name, but the study of poisons also brought about some important advances. Physicians and alchemists including Paracelsus (see pp. 52–53) linked specific chemicals to specific effects on the body and shed light on important concepts of toxicology—such as the relationship between the amount of a chemical and the body's response to it. Through manufacturing, isolating, and analyzing poisons, Renaissance proto-scientists helped lay the foundations for analytical chemistry—the branch of chemical science concerned with detecting and identifying chemicals.

VINEGAR CURE

Ancient texts of the Sumerian and Akkadian civilizations in Mesopotamia recorded knowledge of poisons and emphasized the use of vinegar to detoxify poisons, a practice that continued for thousands of years. We now know that acetic acid in vinegar can effectively attack the chemical bonds that hold poison molecules together, breaking them up into less toxic components.

The earliest known case of poisoning is probably that of Ötzi the Iceman, whose frozen body was found in an Italian glacier 5,000 years after his death.

Renaissance Alchemy

As classical learning filtered into Europe via the Islamic scholars, alchemy became the essential pursuit for men who wanted to understand the hidden workings of the universe. European scholars of the late medieval and early modern period were driven by the quest to read what they called the "book of nature." To them, alchemy seemed to be the key to deciphering this great wisdom.

Magic and Logic

Alchemy can be understood on many levels. The most obvious of its goals was the transmutation of base metals into gold, which alchemists hoped to achieve using a mythical substance known as the Philosopher's Stone. Essentially a sort of magical catalyst, this had many powers attributed to it, and alchemical works such as the *Emerald Tablet* used mysterious words to describe how to make it.

Modern science generally regards the quest for this magical substance as a fantasy pursued by charlatans. It's certainly true that many of those who practiced alchemy were fools motivated by greed, or con artists trying to cheat money from unwise patrons. These men gave alchemy such a bad name that for a time it was banned by kings and popes.

Yet the greatest minds of the era also pursued alchemy. It seemed to offer a rational system for discovering the secrets of nature, and what seems magical to us was in fact considered a form of technology—applying knowledge to allow control of nature. The motivation behind alchemy was basically the same one that created science—the belief that interrogating nature through experiment would allow humans to understand and master it.

A scene from an **alchemist's laboratory**.

Fruitful Searches

The pursuit of the Philosopher's Stone led Renaissance alchemists to make many important discoveries. In 1250, for instance, Albertus Magnus (ca. 1200–1280) isolated arsenic, while his student Roger Bacon (ca. 1214–1292) may have invented black powder (a form of gunpowder). Bacon also recommended experimentation as the best way of uncovering the truth about nature, inspiring later scholars such as Robert Boyle (see pp. 62–63).

The scientist and philosopher **Roger Bacon**. Known as the "wondrous doctor," he may have also invented a type of gunpowder.

A 14th-century alchemist writing under the pen name of "Geber," the European name for Jabir (pp. 42–43), made important discoveries such as vitriol and *aqua fortis*. Vitriol is sulfuric acid, described as the most significant chemical advance since the discovery of iron smelting, while *aqua fortis* is strong nitric acid. These vital tools would later allow individual elements to be isolated from compounds.

Meanwhile the search for the Elixir of Life, another mythical substance supposed to cure all ills and grant immortality, led to breakthroughs in medicinal alchemy, such as the discoveries of Paracelsus (see pp. 52–53). The physician Arnold of Villanova (ca. 1238–ca. 1310), believing that grapes absorbed the essence of the Sun and therefore of gold, distilled wine to produce *aqua vitae*—almost pure alcohol, another important tool for later chemists. Alcohol, like the strong acids, can act as a solvent for substances that don't dissolve in water.

"Without the Philosopher's Stone, chemistry would not be what it is today. In order to discover that no such thing existed, it was necessary to ransack and analyze every substance known on Earth."

—**Justus von Liebig**

Solvents and Solutions

For the alchemists, the process of dissolving a substance seemed almost magical. By using the right liquid, almost any substance could be made to disappear—or so it seemed. Then through evaporation, condensation, or precipitation, the original substance or a new one could be recovered. The core process is now better understood, of course, forming one of the basic concepts of chemistry.

What is a Solution?

A solution is a homogeneous mixture—in other words, a mixture that is the same throughout. This is different from a suspension, where particles of one substance float around in another and can be filtered out. A solution is made up of a solvent and one or more solutes, and generally the solvent is the substance that is in the majority.

Solvents are usually liquids, while solutes can be in any phase (gas, liquid, or solid), but there can also be gaseous and solid solutions. Gases can mix to become homogeneous. The air is a good example—sample it anywhere at sea level and it will have the same gases in the same proportions. The majority of the atmosphere is made up of nitrogen, so this is considered the solvent, while oxygen, carbon dioxide, and so on, are solutes. Solid solutions include metal alloys—bronze, for instance, is a solid solution of tin dissolved in copper solvent.

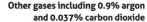

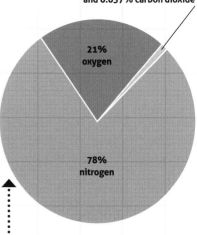

Other gases including 0.9% argon and 0.037% carbon dioxide

21% oxygen

78% nitrogen

The composition of the **Earth's atmosphere.**

Like Dissolves Like

Water is the best-known and most common solvent, but not everything will dissolve in water. The principle that controls solubility is "like dissolves like," where "like" refers to polarity. Polarity is an electrical property found in some molecules, caused by the type of bonds between the atoms.

In a molecule like water, where two hydrogen atoms are bonded to an oxygen atom, the electrons are unevenly distributed so that the oxygen atom has a partial negative charge and the hydrogen atoms have a partial positive charge. This means the molecule itself ends up with a negative pole and a positive pole, like a tiny magnet, and is known as a dipole. Because of its polar nature, water can dissolve other polar solutes, such as salts, sugars, and alcohols. But nonpolar solutes, such as oils, won't dissolve in water, but they will dissolve in nonpolar solvents—for instance, olive oil will dissolve in petroleum.

Solubility and Saturation

The maximum amount of solute that will dissolve in a solvent is described as its "solubility." This is usually measured in grams of solute per 100ml of solvent (g/100ml). For solids, solubility increases with temperature—for instance, you can dissolve more sugar in a cup of hot tea than in a glass of iced tea. For gases dissolving in liquids, this relationship is reversed—as temperature rises, less gas will dissolve. When the maximum amount of solute is dissolved, the solution is said to be "saturated." Sometimes it's possible to dissolve more than the maximum, in which case the solution is said to be "supersaturated."

The concentration of a solution is a measure of how much solute is dissolved in it. Units of concentration include:

Molarity: the number of moles of solute per liter of solution (see pp. 102–103 for an explanation of moles).

Parts-per-million (or billion): commonly used for gaseous solutions.

Percentage: this can be defined by weight, volume, or a combination of both. For instance, in 100g of a salt solution that has a concentration of 10% by weight, there will be 10g of salt.

A

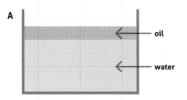

← oil

← water

B

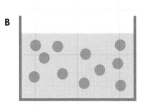

C

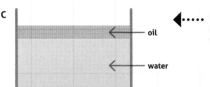

← oil

← water

Oil is not soluble in water (A). If the two are mixed together (B) then left to settle, the oil will rise to the top and form a layer (C).

Paracelsus

One of the more colorful and controversial characters of his era, the 15th-century physician and alchemist known as Paracelsus made significant advances in chemistry, especially in the chemistry of medicine. As importantly, his outspoken manner helped free natural philosophers from the chains of tradition, opening doors to new methods of inquiry and scientific exploration.

Wandering Scholar

Philippus Aureolus Theophrastus Bombastus von Hohenheim was born in Switzerland in 1493, the son of a doctor. Educated at a school that specialized in training engineers for the nearby silver mines, he gained a firm grounding in mineralogy and metallurgy. He then became a traveling scholar, studying medicine at the University of Vienna, Austria, among others, then working as a military surgeon. According to legend, he trained with mystics, alchemists, and physicians in Egypt, Arabia, and the Holy Land.

He later returned to Basel and made a name for himself by curing the printer Froben of a leg infection, though other physicians had advised amputation. He was duly appointed city physician, though didn't keep the post long. Not only did he publicly burn the works of the great physicians Galen and Avicenna, but he also adopted the name Paracelsus—so declaring that he had moved "beyond Celsus," an important Roman medical writer.

After many years' traveling and working in Europe, he returned to his home town in 1541 and was appointed physician to the Duke of Bavaria. Paracelsus died that same year.

Challenging Authority

Paracelsus is chiefly known for his bold new approach to medicine— most notably bringing chemical knowledge to medicine. He understood that specific chemicals have specific effects, and these effects depend on dose. He originated mercury treatments for syphilis, described the anesthetic

ether and concocted laudanum (a tincture of opium) as a general pain-reliever and cure-all. More generally he is credited with helping to set alchemy on the road to becoming chemistry. He developed a new theory of matter, arguing that all substances were composed from sulfur, mercury, and salt—the *tria prima* (three primes)—which represented combustibility, liquidity, and solidity. He even devised a proto-scientific naming system for chemicals, and tried to classify substances according to their chemistry.

But perhaps his true legacy to later philosophers was his willingness to challenge received wisdom. By openly attacking the ideas of scholasticism—the dominant method of teaching based on Aristotle, logic and authority—Paracelsus paved the way for the next generation.

> "Alchemy is the art that separates what is useful from what is not by transforming it into its ultimate matter and essence . . . [it] is an explanation of the properties of all the four elements—that is to say, of the whole cosmos."

RECIPE FOR MAKING A HUMAN

One of Paracelsus' more unusual claims was that he had successfully created a homunculus, or "little man"—an artificial being. According to the recipe he gave, if sperm, "enclosed in a hermetically sealed glass, is buried in horse manure for forty days, and properly magnetized, it begins to live and move. After such a time it bears the form and resemblance of a human being, but it will be transparent and without a body." If fed with an extract of blood it would become a tiny human child that could be raised as normal.

An illustration from **Goethe's *Faust***, showing a student in his alchemical laboratory, trying to create his own homunculus.

Weights and Measures

**Paracelsus famously said, "All substances are poisons . . .
Only the dose determines that a thing is not a poison." In
identifying what modern toxicologists call the "dose-
response relationship," he was among the first to recognize
what would become one of the most important, if least
exciting, elements of the new chemistry that was about to
dawn: the vital need to measure things properly.**

Quantity Not Quality

Aristotelian natural philosophy
was largely qualitative—that is, it
discussed categories, classifications,
qualities, and essences. Although
alchemy introduced some
quantitative thinking, such as
occasional directions on quantities
in the recipes given in alchemical
manuscripts, it too remained mainly
qualitative. The precise amounts of
substances used by alchemists were
less important than their nature.
But chemistry is an extremely
quantitative science, particularly
when it comes to the search for new
compounds and elements. Without
precise measurements of reactants,
there was no way for future chemists
to understand properly the nature of
the products. The need for a new
approach based on measurement
can be traced back to a few
important figures.

Nicholas of Cusa (1401–1464) was a
theologian and natural philosopher.
He argued that it was only through
mathematics that the true nature of
things could be understood.
Nicholas applied this principle to
experiments. He weighed a ball of
wool at different times and noted that

A NEW KILOGRAM?

The kilogram is a golfball-sized
cylinder of platinum and iridium,
locked in a safe in Sèvres, France.
Known as the International Prototype
Kilogram (IPK), it has been used as
the official definition of a kilogram
since 1889. Unfortunately, over time
the IPK has been losing mass, making
it less accurate. However,
metrologists (measurement
scientists) have developed a new
method for measuring a kilogram,
by calculating the exact number of
atoms in a kilogram of silicon, which
may eventually replace the cylinder.

it changed according to how much water was absorbed from the air, so that it became an instrument for measuring atmospheric humidity. He used weight to measure the volume of water in containers, enabling him to estimate pi to a high level of precision. Most famously, he weighed a plant growing in a pot with a degree of accuracy not attempted previously, allowing him to show that it was gaining weight, if only by minuscule amounts. Here was the first inkling that plants could be taking something from the air, and that the air itself had weight.

Objective Measures

The theme of measurement was also stressed by Galileo (1564–1642), who distinguished between primary qualities that could be measured objectively—and so could be shown to be objectively true by experimentation—as opposed to secondary qualities that were perceived subjectively. This distinction was essential to the emerging scientific method (see pp. 66–67). The philosopher Francis Bacon (1561–1626) warned, "God forbid we should give out a dream of our own imagination for a pattern of the world."

PRECISION INSTRUMENTS

According to some, the crucible of modern science was Louvain, in modern-day Belgium. It was here that the mathematician and astronomer Gemma Frisius (1508-1555) and his pupil Gerard Mercator (1512-1594) started making scientific, surveying, and mapmaking instruments. Although the academic establishment still sneered at vulgar "artificers and mechanicks," the instruments of Louvain allowed natural philosophers to measure the world as it actually was, rather than as the ancient texts claimed. Creating links with booksellers and merchants, the Louvain workshops started an international trade that would supply telescopes and microscopes, balances, and scales—literally the instruments of the scientific revolution.

"Measure what can be measured and make measurable what cannot."

—**Galileo Galilei**

Galileo was instrumental in enforcing the distinction between primary and secondary qualities in experimentation.

Pneumatic Chemistry

With his experimental look at air, Nicholas of Cusa had opened a whole new chapter in chemistry—the study of pneumatics (from the Greek *pneuma*, "breath"). Although alchemists had been aware of vapors and airs arising from their crucibles and flasks, little attention had been paid to the world of gases. This was to change, with major consequences for the new science of chemistry.

Gases in the Spotlight

As late as 1727, in his pioneering work on pneumatics, the English scientist Stephen Hales (1677–1761) proposed that gases should now be considered among the chemical principles, even though they had so far been rejected. Gases were the poor relations of the chemical world, ignored and misidentified by alchemists and natural philosophers, who assumed that all "airs" were ethereal and unknowable.

But in the 17th century this began to change, first with the work of Jan Baptista van Helmont (see pp. 58–59), then with a series of dramatic experiments proving the existence of the vacuum. The first of these was the barometer experiment of Evangelista Torricelli (1608–1647, pictured left), performed in 1644.

Torricelli filled a glass tube with mercury, then put his finger over the open end, turned it upside down and lowered the mouth into a basin of mercury. When he took his finger away, the column of mercury inside the tube dropped part of the way and then stopped. Torricelli insisted that the space at the top of the tube was empty—it was a true vacuum. What's more, the column of mercury was being held in place by the pressure of the air on the basin of mercury—in other words, air had weight.

The Mass of Air

These experiments showed that air had mass, while a vacuum was true empty space. Natural philosophers such as Boyle (see pp. 62–63) saw this as clear evidence for the atomic theory of matter, with substances made up of tiny particles separated by void. Eventually such thinking would give rise to a model of how gases behave, known as the "kinetic theory of gases," which describes the properties of an ideal gas.

IDEAL GAS LAWS

The kinetic theory of gases describes the properties of an ideal gas. These include:

- Gases are composed of minute particles. It makes no difference whether these are atoms or molecules, they behave the same. The particles are so small in relation to the distances between them that they take up no volume, which means they can be compressed (unlike a liquid or solid).

- The gas particles move about randomly, in straight lines, until they collide with the walls of their container. These collisions form the pressure exerted by a gas. This constant, random motion allows gases to mix uniformly.

- Gas particles have no forces of attraction or repulsion between them, so they can be considered as completely independent, like minute ball bearings careering around.

- The average kinetic energy of the gas particles determines the temperature of the gas.

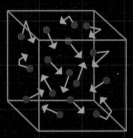

Gas particles move about randomly, in straight lines, until they collide with the solid walls of their container.

THE POWER OF AIR PRESSURE

An even more dramatic display of a vacuum followed in 1654. The German engineer Otto von Guericke (1602-1686) had modified a water pump to create the first air pump. In front of the Emperor Ferdinand III, two giant copper hemispheres were fitted together and the air pumped out of them. Although nothing held them together, the hemispheres couldn't be pulled apart by 16 horses. When Guericke released a valve to readmit the air, the spheres fell apart by themselves.

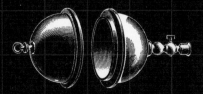

Jan Baptista van Helmont

The father of pneumatic chemistry was a reclusive Flemish nobleman with a mystical disposition and arcane beliefs. Yet he is credited with the first controlled experiment in biochemistry and he anticipated important developments and laws in chemistry. His groundbreaking experiments led him to conclusions that were centuries ahead of his time.

Solitary Researcher

The alchemist and physician Jan Baptista van Helmont was born into a noble family in Brussels, Belgium, in 1579. After studying at Louvain University and traveling around Europe, he retired to his country estate to pursue mystical and scientific researches. Despite being very religious, he ran into trouble with the Catholic Church because of his involvement in a controversy over the powder of sympathy (see box). After insisting there was nothing magical about it, he was put under house arrest. Not until after his death in 1644 was his son able to publish his collected writings as *Ortus Medicinae* (*Origin of Medicine*) in 1648.

Watery Origins

Van Helmont's most famous research was an improved version of Nicholas of Cusa's experiment on plant growth (see pp. 54–55). He weighed a willow tree and some dried soil, then planted the tree. He covered the pot and fed the plant distilled water. After five years he reweighed the tree, which had increased in mass by 169lb (76kg). He dried and reweighed the soil and found it was almost the same mass. Van Helmont concluded that the tree grew by drinking water. Linking this with findings from other research, he saw this as proof that matter largely consisted of water (just as Thales had claimed 2,000 years earlier).

Spirits of the Air

Van Helmont had overlooked the role of carbon dioxide in plant growth, but he was the first to suggest its existence, after another innovative experiment involving careful weighing. After setting fire to 62lb (28kg) of charcoal he found that only 1lb (50g) of ash was left. He had previously demonstrated that matter could not be destroyed, only changed in form (anticipating "the law of conservation of mass" by over a century). He then surmised that the other 61lb (27.5kg) of matter had escaped as a form of vapor, which he named "gas" after the Greek word for chaos.

The gas from burning charcoal he named *spiritus sylvester* ("spirit of the wood"). In other experiments in combustion he identified a second form of *spiritus sylvester*, gas carbonum and gas pingue. These four gases are now called carbon dioxide, carbon monoxide, nitrous oxide, and methane.

THE POWDER OF SYMPATHY

One of the magical principles of alchemy is the power of "sympathy"— the belief that objects or substances, once associated, continue to have influence over one another. This principle was the reasoning behind Paracelsus' peculiar belief, later taken up by van Helmont. He theorized that a special ointment could treat wounds by being applied to the blade that had inflicted them, The ingredients of this ointment, later known as the "powder of sympathy," included moss from the skull of a man who had died a violent death, boar and bear fat from animals killed while mating, burnt worms, dried boar's brain, red sandalwood, and powdered mummy.

> "I call this Spirit, unknown hitherto, by the new name of Gas, which can neither be constrained by Vessels, nor reduced into a visible body." —**Jan Baptista van Helmont**

Acids and Bases

One of the oldest distinctions in chemistry was between acids and alkalis. Acids had sharp, sour tastes and were able to "dissolve" earths and metals; alkalis had a bitter flavor and often a soapy feel. A third class of substance was known as salts, but by the 17th century it was becoming clear that salts could be the product of "opposition" between acids and alkalis, and substances that reacted with acids to form salts (including metals) were called "bases."

Chemical Conflict

Acids and their "opponents" the bases were to play a vital role in chemistry becoming a science. Since the dawn of antiquity, alkalis had been known in the form of soda (sodium carbonate) and potash (potassium carbonate). Known as "fixed alkalis" because they were nonvolatile, these contrasted with volatile alkalis, such as ammonia. To these were later added "alkaline earths," the name given to calcium carbonates from chalk and limestone, and later to salts of magnesium and other metals.

Meanwhile organic acids such as vinegar and lemon juice had been known since ancient times. To these Islamic and medieval European alchemists added spirit of vinegar (purified acetic acid) and inorganic acids such as spirit of salt (hydrochloric acid), which were much more powerful.

Their reactions with the alkali could be quite violent, with effervescence and heat. But explaining how they worked and what made them acidic or alkaline proved difficult until Robert Boyle (see pp. 62–63) discovered a way of classifying them using plant infusions—an early version of today's litmus test. He found that syrup of violets was blue in its natural state but turned red with acid and green with alkali.

Evolving Theories

Explanations of how acids and bases work have evolved over time. First came alchemical ideas about opposing "male" and "female" principles, followed by theories linking acidity to phlogiston (see pp. 72–73). Oxygen was seen to cause acidity until Humphrey Davy showed

The Swedish chemist **Svante Arrhenius** devised the first modern definition of acids and bases.

that hydrochloric acid (HCl) lacked oxygen, proposing instead that hydrogen caused acidity. Then the Swedish chemist Svante Arrhenius (1859–1927) defined an acid as a substance that produces a hydrogen ion (a proton, H^+), and an alkali as one that dissolves to produce a hydroxide ion (OH^-), as shown in the chemical equations for a) hydrochloric acid and b) sodium hydroxide:

a) $HCl(aq) \rightarrow H^+ + Cl^-$

b) $NaOH(aq) \rightarrow Na^+ + OH^-$

(Note that "aq" denotes "aqueous solution," a solution in which the solvent is water.) Here, reactions between acids and bases are "neutralization" reactions, because they produce water and a neutral salt:

$HCl(aq) + NaOH(aq) \rightarrow$
$H_2O(l) + NaCl(aq)$

The water is produced when the hydrogen ion from the acid and the hydroxide ion from the base get together.

Arrhenius's model is accurate for acids and bases that are aqueous solutions (containing water), but it's also possible to have acid-base reactions between gases, and so a more general theory of acids and bases was needed. Known as the "Brønsted–Lowry

ACIDS AND BASES IN EVERYDAY LIFE

There are probably plenty of acids and bases, or alkalis, around your home. Here are some of the typical ones:

Acids: vinegar (acetic acid); carbonic acid (found in sodas and carbonated water, formed when bubbles of carbon dioxide dissolve in water); acetylsalicylic acid (also known as aspirin); sulfuric acid (found in car batteries).

Alkalis: ammonia, used as a cleaner; lye (sodium hydroxide), another cleaner; baking soda (sodium bicarbonate); stomach-settling antacids such as calcium carbonate and aluminum hydroxide.

Acid-Base Definitions

Type	Acid	Base
Arrhenius	H^+ producer	OH^- producer
Brønsted-Lowry	proton (H^+) donor	proton (H^+) acceptor

theory," it sees acids as proton donors and bases as proton acceptors. In the Arrhenius model, the H^+ is a proton donor and the OH^- a proton acceptor.

Robert Boyle

The centuries of alchemical discoveries seemed to be building toward something: a break with the past and a new, scientific approach to chemistry. This pioneering shift was personified in the life and career of Robert Boyle, an Anglo-Irish nobleman whose discoveries in experimental and pneumatic chemistry led to him being regarded as the "father of scientific chemistry."

"Chymists" and Philosophers

The fourteenth son of a wealthy earl, Robert Boyle (1627–1691) was expensively educated and traveled Europe as a youth. He concentrated on theological studies in his early years before meeting a circle of alchemists and natural philosophers. In particular he became involved with the American alchemist George Starkey (1628–1665), who schooled him in alchemy, teaching him the skills of the "chymist." At this time "chymistry" was seen as a suspect pursuit, mixing the unsophisticated skills of the artisan (such as an apothecary) with the mystical quest for the Philosopher's Stone and the transmutation of gold.

In the 1650s Boyle moved to Oxford, England, where he made experimental breakthroughs as he pursued his lifelong goal: to marry the expertise of the practical "chymists" with the loftier ambitions of natural philosophy, which aimed to understand the universe. He later moved to London and was a founder member of the Royal Society, the scientific academy seen as the crucible of the scientific revolution.

A New Approach

Boyle's chemical discoveries included a color test for acids, numerous medical treatments, and extensive work with air pumps and vacuums, which led him to formulate what is now known as Boyle's law: that the pressure of a gas is inversely proportional to its volume. In other words, if you compress a volume of gas by half, its pressure will double.

Boyle supported the atomist philosophy, though he preferred the term "corpuscles" to "atoms." For him, the corpuscularian theory represented a break with pre-

P = 100 kPa (0.987 atm)

v = 6 dm³ (6 liters)

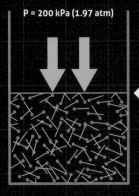

P = 200 kPa (1.97 atm)

v = 3 dm³ (3 liters)

As **gas is compressed** into a smaller volume, so its pressure (measured in kPa, kilopascals) increases.

scientific Aristotelian chemistry, and to advance and defend this new theory he used experiments. For instance, he demonstrated how the chemistry of saltpeter (a component of gunpowder) could be explained simply in terms of the size and motion of corpuscles, without the need for any explanations to do with "forms" and "qualities."

It was this effort to replace the old way of thinking that led to Boyle's most famous work *The Sceptical Chymist*. It attacked the doctrine of four elements and Paracelsus's *tria prima* (see pp. 52–53). And it tried to persuade "chymists" to adopt a more philosophical approach to the study of nature. It was this approach, rather than his actual discoveries, that secured Boyle's reputation as the "father of scientific chemistry."

A WISH LIST

In the 1660s Boyle put together a series of notes listing the most urgent problems for scientists to solve. These included:

• The secret of eternal youth.
• Curing diseases by transplant.
• Developing pain-relief pills.
• Perfecting flying.
• Finding ways for people to work underwater.

Curiously he also recommended research aimed at "Attaining Gigantik Dimensions," believed to be a reference to the possibility of enlarging the human race.

Ionic and Covalent Bonds

Before going any further with the development of scientific chemistry, it's time to introduce the concept of chemical bonding. Let's look at the two main types of bonds and the underlying principle of chemical bonding, which is the tendency for electrons to distribute themselves in space around atoms so as to lower the total energy of the group.

The Octet Rule

When atoms join up with other atoms to form compounds they like to find a configuration with the lowest possible energy. If the total energy level of a group of atoms is lower than the sum of the energies of the individual atoms, they bond together and become lower in energy.

An atom's energy configuration is mainly controlled by the distribution of its electrons. Most importantly for chemical bonding, the outermost shell of electrons orbiting an atom (see pp. 28–29) is known as the valence shell, and its completeness determines bond formation and the reactivity of that atom.

Valence shells follow the octet rule, which states that the most stable, lowest-energy configuration is for the shell to have eight electrons. The elements that already have eight electrons in their outermost shells are the noble gases, including helium, neon, and argon. Also known as the inert gases, these are very unreactive, as they have stable, full valence shells. As a general rule in chemical bonding, electrons will try to transfer themselves to achieve the valence shell configuration of the nearest noble gas. The ability of electrons to transfer is the key to bond formation.

As a general rule in chemical bonding, electrons will try to transfer themselves to achieve the valence shell configuration of the nearest noble gas.

Give and Take

Chemists discovered the two types of chemical bond, ionic and covalent, when testing solutions for their electrolysis properties (see pp. 112–113). Some compounds dissolve in water to produce conductive solutions, known as electrolytes, while others don't—these solutions are known as non-electrolytes.

In an **ionic bond** one atom transfers one or more electrons to another atom. The donating atom sheds its "incomplete" outermost electron shell, so that the "complete" shell beneath becomes the new valence shell. The receiving atom fills up its outer shell so that it becomes complete. For instance, when sodium and chlorine atoms bond to form sodium chloride (NaCl), or table salt, the sodium atom loses one electron to achieve a neon-like arrangement of its electrons; the chlorine atom gains an electron to achieve an argon-like arrangement. As a result, the atoms become ions—a sodium ion that is positively charged (a cation), and a chlorine ion that is negatively charged (an anion). So the more accurate chemical formula for table salt is Na^+Cl^-. There's an electrostatic attraction between positive and negative ions, and this binds the particles together in an ionic compound.

COMPOUNDS CHECKLIST

Ionic compounds	Covalent compounds
Electrolytes ☑	Non-electrolytes ☑
Generally solid at room temperature ☑	Can be solids, liquids, or gases ☑
Higher melting point ☑	Lower melting point ☑

In a **covalent bond** two atoms share a pair of electrons—the electrons basically take up a new orbit encompassing both atoms. For example, bromine exists in nature as a diatom (Br_2) because it's seeking to gain a full krypton-like valence shell. A single bromine atom has seven electrons in its outermost or valence shell; to become like krypton, it needs eight, so two bromine atoms share an electron pair, allowing each to fill its octet and achieve a stable, low-energy configuration.

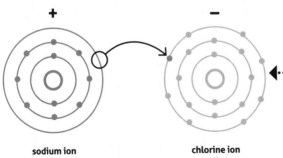

sodium ion
Na+

chlorine ion
Cl⁻

The **electron configurations** of a sodium ion and a chlorine ion, showing how an electron is donated to allow both to achieve a valence shell like that of a noble gas.

The Scientific Method

What does it mean to say that chemistry before Robert Boyle was "unscientific" or "proto-scientific," but that during the 17th century it became a science? What was so different about the chemistry of Jabir or Paracelsus and that of Boyle? The answer lies in a new methodology and a new philosophy, brought together to give a unified system of incredible power—the scientific method.

The Trouble with Alchemy

We've already touched on some of the unscientific characteristics of alchemy, but it's worth explaining them in detail to highlight the contrast with what came after. At its heart, alchemy is based on statements and theories held to be true, even though they haven't been tested or proved. For example, the assumption that there are four basic elements, or the belief that there are connections between metals and signs of the zodiac. Alchemical procedures, techniques, and recipes put emphasis on factors that are likely to change, such as the mental and spiritual state of the experimenter—so it was believed an experiment might fail because the experimenter didn't have a sufficiently pure spirit, for instance.

Rather than sharing results and the details of experiments, such as techniques and quantities, so that others could examine, critique, and replicate, as in science, alchemists believed in hiding results and techniques through mysterious symbols. Finally, alchemists resisted attempts to formalize their art or bring knowledge together into coherent systems or theories.

No More Speculation

Alchemy wasn't the only troublesome subject studied by natural philosophers. In medicine, astronomy, biology, and physics, the common approach was scholasticism, which was based on assumptions and the importance of authority. From Francis Bacon (1561–1626) to Robert Boyle and his younger colleague Isaac Newton (1642–1727), the new breed of natural philosophers sought to pioneer a new way of doing things. Their agenda was to move away from speculations unsupported by evidence, instead favoring experiments that observed nature as it actually was.

Boyle and Newton were key figures in developing a new, scientific method. In simple terms, the scientific method is that observations of nature—perhaps from an experiment—lead to an initial hypothesis to explain the phenomena. In turn this leads to experiments that test the hypothesis. If the results of the experiments don't support the hypothesis, they must be changed or discarded. (Boyle, for instance, was ahead of his time in recognizing the value of unsuccessful experiments.) If the results *do* support the hypothesis, and they can be replicated, then the hypothesis may become a theory. If repeated patterns can be observed, and better still mathematically quantified, the theory may give rise to laws or general truths. If new evidence comes along that doesn't fit with the theory, the theory must be changed or discarded. According to its supporters, the scientific method is the only sure route to truth.

"We are certainly not to relinquish the evidence of experiments for the sake of dreams and vain fictions of our own devising." — **Isaac Newton**

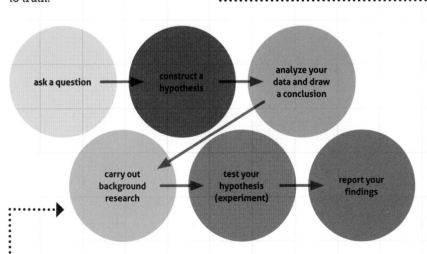

A simplified version of **the scientific method** in action, stressing the need to report the findings of experiments so that others can attempt to replicate and verify them.

3

Tracking Down the Elements

The scientific revolution was the catalyst chemistry had been waiting for, and the new science advanced in leaps and bounds. With new concepts, equipment, and techniques, it was possible to discover new elements at a rate undreamed of by the ancients. Fortune and glory awaited those who could reveal new pages of the book of nature, and this chapter describes the excitement of the race to identify the principles of the new science.

Carl Scheele

In 1669 German alchemist Hennig Brand won fame and fortune by isolating phosphorus from urine. However, it wasn't until the mid-18th century that a flood of further elemental discoveries was unleashed, as chemical analysis became more sophisticated. Sweden was the initial epicenter of these discoveries, in particular Carl Scheele, a man who never got the credit he deserved.

Goblin Ores

Medieval miners knew a lot about metals and ores, but it was mainly superstitious folklore. Two examples were *Kobolds* and *Nickels*, German terms for goblins and gremlins, which were blamed for strange sounds, fumes, and mishaps in the mines. *Kobolds* were held responsible for the toxic fumes of a false copper ore, which gives glass a vibrant blue color when smelted. In 1735 Swedish chemist Georg Brandt (1694–1768) showed that the color was due to a metal in the ore, which he named cobalt. The *Nickel*, or scamp, was blamed for a false copper ore called *Kupfernickel* ("copper scamp"). In 1751 Swedish chemist Axel Cronstedt (1722–1765) revealed that it contained a hard white metal, which he named nickel.

Prolific Discoveries

Carl Wilhelm Scheele (1742–1786) was a largely self-taught chemist from Pomerania, an area in modern-day Germany and Poland. He came from a poor family and received little formal education until he became an apothecary's apprentice. Despite this, Scheele became an expert chemist and fanatical experimenter. After working all over Sweden, he took over a pharmacy in the small town of Köping and remained in this job for the rest of his short life, refusing prestigious academic posts.

The list of Scheele's discoveries is remarkable, spanning organic and inorganic chemistry, and most notably includes oxygen and chlorine. Scheele was a practical man, not a theorizer, and though he may not have recognized or identified some of his discoveries, it is extraordinary that he received such little credit at the time. Largely this was due to a stroke of bad luck. In 1773 he finally completed his book *A Chemical*

Without knowing it, **Scheele was the first to produce oxygen**. When he heated a variety of substances, including potassium nitrate and magnesium dioxide, they gave off the same gas, "fire air." He named it for the sparks produced when it came into contact with charcoal dust.

Treatise on Air and Fire but had to wait four years for it to be published. This meant that his discovery of oxygen, though predating Joseph Priestley's, was only announced three years after the Englishman had published his own findings (see pp. 84–85).

Fire Air

Scheele had discovered oxygen by separating air into two main components, one of which burned easily. Calling it "empyreal air" (fire air), he went on to produce it in a series of experiments, showing that it played a role in plant and fish respiration. Scheele didn't fully understand the result of his finding, explaining its properties using the phlogiston theory of combustion (see pp. 72–73). And when he discovered chlorine in 1774, he mistakenly thought it was a compound of oxygen—the gas wasn't recognized as an element until Humphry Davy's work on hydrochloric acid (see pp. 114–115). Eventually, Scheele's hands-on method of analysis, which involved tasting many of his discoveries, caught up with him and he died at age 43.

> "Since this air is absolutely necessary for the generation of fire, and makes about one-third of our common air, I shall henceforth call it empyreal air [fire air]." **—Carl Scheele**

"PEE" FOR PHOSPHORUS

German scientist Hennig Brand (ca. 1630-1692) was convinced he could distill gold from urine. He stockpiled 60 tubs of urine in his basement laboratory and allowed it to putrefy. Then he boiled it to produce a paste, which he heated and drew into water to condense the vapors, ending up with a waxy white substance that glowed in the dark. He named it phosphorus, after the Greek for "bringer of light."

The Phlogiston Theory

Phlogiston was a hypothetical substance believed to be the main agent involved in combustion, reduction, and respiration. Although now ridiculed as a scientific dead-end that led serious chemists astray, this landmark in the history of chemistry was also a valuable theory and a useful illustration of the power of the scientific method.

A Burning Secret

Alchemists and early chemists were fascinated by combustion, rusting, respiration, fermentation, and calcination (heating a substance to a high temperature but still below its melting point). These processes were clearly interlinked, and uncovering the hidden relationships between them—especially the common principle— promised to reveal one of nature's deepest secrets. We now know that the common principle is oxygen, and that, for instance, combustion of wood is a form of oxidation where carbon is oxidized to carbon dioxide, leaving only ash behind; while what early chemists called a calx—the powdery substance left behind when a pure metal is heated in air—is a metal oxide. But before oxygen had been discovered, there were several other sensible hypotheses.

German scientist Joachim Becher (1635–1681) argued that combustible substances contained an active agent called *terra pinguis* ("fat

The German chemist **Georg Ernst Stahl**. His theory of phlogiston dominated European chemistry until the chemical revolution at the end of the 18th century.

earth"). Jan Baptista van Helmont adopted this hypothesis and coined the term "phlogiston," from the Greek for "inflammable." But it was the German chemist Georg Ernst Stahl (1660–1734) who first offered a proper theory of phlogiston. He saw that burning a piece of charcoal produced flames and smoke, and left behind a small amount of ash. To him, the heat and smoke represented something being driven off, and that something was the agent of combustion—phlogiston. He deduced that charcoal is composed of ash and phlogiston. A reversal of the process is the reduction of the combustion products—as when heating a calx with charcoal to produce pure metal. The calx absorbs phlogiston from the charcoal to produce the metal.

Flawed Theory

As the first rational explanation for combustion and calcination, phlogiston made perfect scientific sense. The French chemist Pierre-Joseph Macquer believed it had "changed the face of chemistry" and urged chemists to search for new

substances that would prove its existence. Joseph Priestley adapted the theory to explain the gases he'd discovered (see pp. 84–85). But as chemistry became more precise, the theory ran into trouble.

Stahl's theory suggested that combustion should result in loss of mass as phlogiston is driven off, while reduction of calx with charcoal should add mass as phlogiston is absorbed—but experimental results showed the opposite. (We now know that metal heated in air gains mass because it has combined with oxygen, whereas reduced calx loses mass because it loses oxygen.) Stahl and other phlogiston supporters tried to argue around this, but phlogiston was finally killed off in 1774 by Lavoisier's oxygen theory (see pp. 86–87).

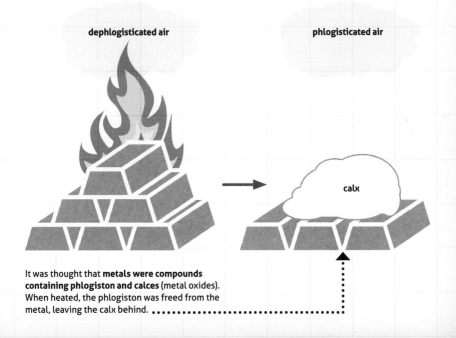

dephlogisticated air

phlogisticated air

calx

It was thought that **metals were compounds containing phlogiston and calces** (metal oxides). When heated, the phlogiston was freed from the metal, leaving the calx behind.

Carbon Dioxide

Pneumatic chemistry was gathering pace as increasingly sensitive instruments and sophisticated techniques allowed the analysis of gases, or "airs" as they were known. Alchemists such as van Helmont had been able to guess the existence of different types of airs in a qualitative way, but the new breed of scientists were able to prove their existence using quantitative measures.

Fixed Air

From 1754 to 1756, the Scottish chemist Joseph Black presented an impressive series of experiments where he obtained an unknown gas from heating chalk (calcium carbonate) to produce quicklime (calcium oxide). As it was contained within the solid until released by heating, he called the new gas "fixed air." It was the same gas described as *spiritus sylvester* by van Helmont a century earlier—what we now call carbon dioxide (CO_2).

Joseph Black (1728–1799) proved that carbon dioxide was involved in the processes of life— including breathing, photosynthesis, and fermentation. He later made breakthroughs in the field of heat (see pp. 82–83).

Black went on to demonstrate a complete cycle of chemical transformations of this new air. Having decomposed chalk to produce quicklime, he showed he could reverse the process and recombine the fixed air with quicklime to produce chalk. He then proved that the same fixed air, which he identified by weighing it, was the product of combustion, fermentation, and respiration. Although he didn't pursue research into fixed air, he correctly surmised that it formed a component of the atmosphere (CO_2 makes up around 0.037% of the air).

Black's experiments with fixed air were part of his research into "causticization"— the opposite of acidification (making something more acidic). He described how carbonates, which he defined as mild alkalis, are causticized—made more strongly alkaline—when they lose fixed air. Whereas when carbonates take up fixed air, they are reconverted into mild alkalis. Black also demonstrated how release of carbon dioxide caused effervescence when limestone was added to acids.

Breath of Life
....

In a dramatic demonstration of his claim that fixed air was the product of respiration, Black exhaled into a jar of limewater (a solution of calcium hydroxide, also called slaked lime), which turned cloudy as tiny particles of chalk formed within. This is still the standard test for the presence of carbon dioxide:

$$CO_2(g) + Ca(OH)_2(aq) \rightarrow CaCO_3(s) + H_2O(l)$$

If you continue to bubble carbon dioxide through the mixture, it will turn clear again, as calcium carbonate and carbon dioxide react together to form calcium hydrogen carbonate—a colorless solution:

$$CO_2(g) + CaCO_3(s) + H_2O(l) \rightarrow Ca(HCO_3)_2(aq)$$

Hardness of water is caused by similar reactions, where rainwater is acidified by carbon dioxide and then reacts with limestone in the ground.

GREENHOUSE GAS

Carbon dioxide is one of several "greenhouse gases" in the atmosphere that trap heat. Sunlight (short-wave radiation) passes through the atmosphere, and the Earth's surface absorbs part of this energy and heats up. The Earth cools down by giving off long-wave infrared radiation, but before this can escape to space, some of it is absorbed by greenhouse gases. This makes the atmosphere warmer, which in turn makes the Earth's surface warmer. The greenhouse effect keeps the planet's temperature around 95°F (35°C) higher than it would otherwise be, making life on Earth possible. But an increased greenhouse effect makes the Earth warmer than normal, leading to global warming and climate change.

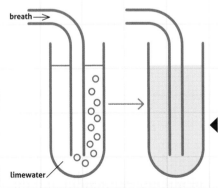

breath →

limewater

In order to demonstrate that **CO₂ is a product of respiration**, Joseph Black breathed into a jar of limewater (a solution of calcium hydroxide), which turned cloudy as tiny particles of chalk (calcium carbonate) formed within. ....

Henry Cavendish

The man who made the greatest strides in the field of pneumatic chemistry was an eccentric English millionaire, generally regarded as the greatest man of science since Isaac Newton but pathologically shy and retiring. Henry Cavendish unlocked the secrets of the atmosphere and created water, but couldn't bring himself to speak to or even look at a woman.

Airs and Graces

Joseph Black's experiments (see pp. 74–75) inspired English aristocrat Henry Cavendish (1731–1810) to begin his own research into airs. Cavendish was the grandson of the Dukes of Devonshire and Kent and destined to become one of the richest men in Britain when his father died. However, his only interests were scientific, and he was a noted eccentric and recluse (see box).

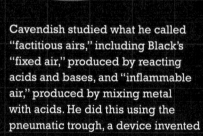

Cavendish studied what he called "factitious airs," including Black's "fixed air," produced by reacting acids and bases, and "inflammable air," produced by mixing metal with acids. He did this using the pneumatic trough, a device invented

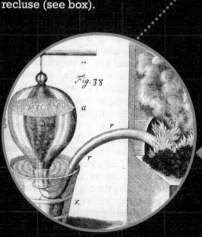

Cavendish used his **pneumatic trough**, shown here in an illustration from 1727, to measure the properties of gases with accuracy.

by Stephen Hales (see pp. 56–57), which captured gas in an upturned vessel over a water tank. By measuring the amount of water displaced, Cavendish was able to calculate the specific gravity of the airs—their density relative to the atmosphere as a whole— showing that his inflammable air was the lightest substance yet discovered.

We now know that Cavendish's inflammable air was hydrogen, but he believed in the phlogiston theory (see pp. 72–73) and thought he might have discovered phlogiston itself. Later he identified the primary components of the atmosphere (also known as "common air"), oxygen and nitrogen, though he called them "dephlogisticated air" and "phlogisticated air" respectively. In 1781 he combined inflammable air with common air and lit the mix with an electric spark, producing water droplets. Measuring the remaining gas, he showed that about a fifth of the common air had vanished. Later he repeated the experiment with just inflammable and dephlogisticated air (hydrogen and oxygen), producing pure water. This finally disproved the ancient concept that water was one of the basic elements (though he wrongly claimed it was the combination of phlogiston and dephlogisticated air that produced water). He even noted that a tiny remnant of common air was inert; a century later this was

A SHY SCIENTIST

Cavendish had an odd personality. According to a fellow of the Royal Society, he never appeared in London unless lying back in his carriage and he seldom spoke. He was so shy that when he forced himself to attend Society meetings, he would give out "shrill cries" as he entered a room of people. If looked at directly he would "retire in great haste," and if spoken to he would flee home. He was especially shy around women, communicating with his housekeeper only through notes and forbidding female servants to approach him. His immense wealth mattered little to him, and even his death was eccentric. Knowing his end was near, he gave strict instructions to be left alone until a time when he had calculated he would be dead. When a concerned butler came in half an hour early, he was sent away with a scolding.

"He was acute, sagacious, and profound, and I think the most accomplished British philosopher of his

Water

Ordinary water is the most familiar liquid to us, yet it's also one of the most unusual, in terms of its chemistry and physical properties, making it the most important substance on Earth. Water occupies a central role in chemistry; known as the universal solvent, it brings about phenomena such as acidity and alkalinity. For life on Earth, it is essential in a host of different ways.

Angular and Polar

The type and distribution of bonds in a compound determine its structure and shape, which in turn determine the properties of the compound. Water is composed of an oxygen atom covalently bonded to two hydrogen atoms, so rather than having a linear shape (H—O—H) it has an angular shape:

Its unusual properties, so vital to life on Earth, are a direct result of this angular shape, which basically gives the molecule sides or "ends." Because oxygen is more electronegative than hydrogen, it more strongly attracts the electron pairs in each covalent bond, so that they're pulled closer to the oxygen atom. This in turn gives the oxygen a partial negative charge, while the hydrogen atoms each have a partial positive charge—so the molecule as a whole has a negative pole and a positive pole, making it a dipole.

Electrostatic Attraction

One consequence of this polarity is that water molecules interact with each other—the partially positively charged hydrogen atom of one molecule is attracted to the partially negatively charged oxygen atom of another. This interaction is called a hydrogen bond. Hydrogen bonding between water molecules gives water many of its unique properties, such as its unusually high boiling point. Generally the boiling point of a liquid relates to its molecular weight (see pp. 102–103), and substances with equivalent molecular weight to

Water's unusual properties, so vital to life on Earth, are a direct result of its angular shape.

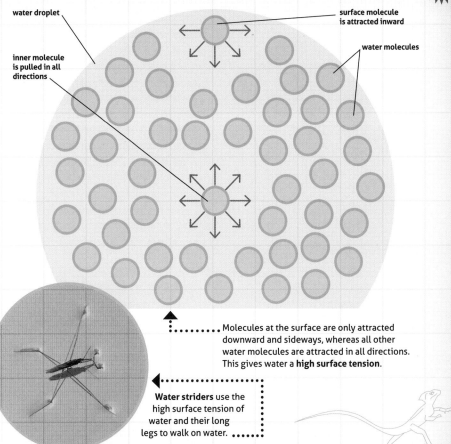

water droplet

surface molecule
is attracted inward

water molecules

inner molecule
is pulled in all
directions

Molecules at the surface are only attracted
downward and sideways, whereas all other
water molecules are attracted in all directions.
This gives water a **high surface tension**.

Water striders use the
high surface tension of
water and their long
legs to walk on water.

FLEXIBLE SURFACE

Water molecules in the liquid phase
are more strongly interlinked and
interactive than most other liquids,
and this is most noticeable at the
surface of a body of liquid water.
Because the molecules at the
surface are only attracted downward
and sideways, whereas all the other
molecules are attracted in all
directions, water has a very high
surface tension. This allows small
creatures such as Jesus lizards to
walk on the surface of water. It
also means that water evaporation
rates are much lower than would be
expected, which helps keep much of
Earth's water in the oceans rather
than in the atmosphere.

water can absorb large amounts of heat and release it slowly; on a global scale, this helps to prevent the kind of wild swings between day and night temperature seen on other planets, and moderate longer-term climate variation. When water freezes, the hydrogen bonds lock the molecules into a rigid matrix that is less dense than the liquid form, which means that ice floats on water instead of sinking. Because of this, only the top of a body of water will freeze, while the rest is insulated.

Dissolving Power

Water's polarity and angular shape make it a powerful solvent for both ionic and polar covalent substances (see pp. 64–65). Its partially charged poles allow it to interact strongly with ions. So when an ionic substance dissolves in water, the water molecules surround the anions (negatively charged ions) with their positive poles, and they surround the cations (positively charged ions) with their negative poles. A similar process allows water to dissolve polar covalent substances. Many organic compounds, such as sugars, alcohols, and proteins, contain O-H and N-H bonds, which are polar, so they will dissolve in water. An ion surrounded by water molecules is known as a hydrate; it is more correct to describe an ion such as Cu^{2+} (ionized copper) as $[Cu(H_2O)_6]^{2+}$. Many inorganic substances can form crystalline hydrate solids.

Ice floats on water because the solid form is less dense than the liquid form.

water are usually gases at room temperature. Water, by contrast, is liquid across a very wide range of temperatures, making it a stable medium for life on Earth.

Hydrogen bonding also gives water a very high heat capacity—the amount of heat needed to change its temperature. And it gives it a high heat of vaporization—the amount of heat energy needed to make it change phase (see pp. 82–83 for more on heat). As a result, bodies of

Liquid water also separates to a small degree, so that $H_2O \rightleftharpoons H^+ + OH^-$. The symbol $\rightleftharpoons$ shows the reaction is reversible and takes place in both directions, so the forward and reverse reactions happen at the same rate. Substances that increase the number of H^+ ions are acidic and those that increase the number of OH^- ions are basic or alkaline.

Water is liquid across a very wide range of temperatures, making it a stable medium for life on Earth.

A LIGHT TOUCH

Water strongly absorbs infrared light but is transparent to visible and near ultraviolet radiation. This means that water vapor in the air lets through solar radiation during the day to heat the planet, but it restricts heat loss at night, maintaining relatively even temperatures across the day-night cycle. Water vapor is also a greenhouse gas (see p. 75).

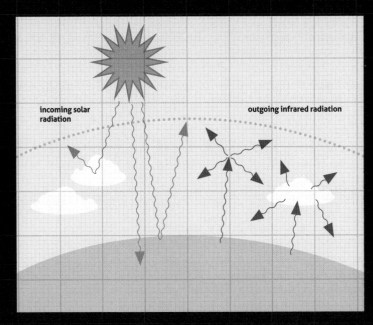

incoming solar radiation

outgoing infrared radiation

Heat

For the ancients, fire had been one of the primal elements, and although the classical elements had been superseded by the mid-18th century, chemists still puzzled over the nature and behavior of heat. Fascinating observations and clever experiments would soon reveal important aspects of the chemistry of heat. In fact, the principles of heat are fundamental to chemistry as a whole.

Hidden Heat

Dutch thermometer pioneer Daniel Gabriel Fahrenheit (1686–1736) had made a curious observation. He noted that the temperature of supercooled water, which immediately changed phase to ice when shaken, shot up to 32°F on his scale. Joseph Black (see pp. 74–75) did his own research and he also observed an apparent mismatch between the heating of water and its temperature. When he melted ice, he noticed that although it absorbed heat, its temperature didn't change—in other words, it went from ice at 32°F (0°C) to water at 32°F. It seemed that the heat had somehow combined with the water particles in such a way that it was "hidden" from the thermometer, and so he called it "latent heat." He was even able to measure this latent heat, which is called the "latent heat of fusion" when a substance changes from a solid to a liquid or a liquid to a solid, and the "latent heat of vaporization" when a substance changes from a liquid to a gas.

Black followed up these discoveries by finding that equal masses of different substances need different quantities of heat to change their temperatures by the same amount. This is known as "specific heat." The specific heat of a substance is the energy needed to raise 1g by 1°C.

A more general concept is "heat capacity," which is the amount of heat required to change the temperature of a substance. Energy is measured in calories or joules, so

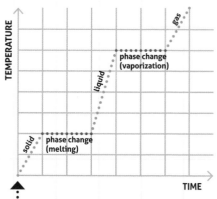

The **temperature of a substance** remains constant while it is changing its state.

SETTING TEMPERATURES

Fahrenheit's temperature scale was the first to be widely adopted. He set the 0° point at the lowest temperature he could achieve, using a mixture of salt and ice. In 1742 Swedish scientist Anders Celsius (1701-1744) proposed that scientific measures of temperature should be made on a fixed scale based on the freezing and boiling of water (at sea level). He suggested 0° as the temperature at which water boils, and 100° as that at which water freezes, but his pupil inverted the scale and it was adopted across Europe as the Celsius scale. Scientists often prefer to use the Kelvin scale, named for Lord Kelvin (1824-1907). This starts at absolute zero, a theoretical state where no energy at all is present. On the Kelvin scale, which is measured in kelvins (K), water freezes at 273K. 1K = 1C° = 1.8°F.

specific heat is measured in calories or joules per g°C. The specific heat of water is 1 calorie/g°C or 4.186 joules/g°C.

A Matter of Misunderstanding

Black's discovery of latent heat fitted well with contemporary beliefs of the nature of heat. If it somehow combined with water particles and was locked away, was it not similar to phlogiston or fixed air? This was how it seemed to 18th-century scientists, who imagined heat as a form of matter (the "matter of fire"), either particles or an elastic fluid. Lavoisier (see pp. 88–89) later formalized this concept of heat by calling it "caloric," but it took another 70 years until chemists came to accept that heat was a form of energy, distinct from matter.

Joseph Black was the first to notice that heat and temperature are not the same thing. Temperature is the average kinetic energy of individual particles in an object, measured in °F, °C, or K. Heat is the amount of thermal energy in an object, measured in joules. For instance, a cup of water and a bathtub full of water may be the same temperature. But the bath contains much more heat than the cup, as it holds more water and so stores more thermal energy.

Joseph Priestley

In the late 18th century, chemistry started to become extremely popular, its discoveries and rivalries matters of national interest. The man partly responsible for this was Christian minister and political radical Joseph Priestley, inventor of soda water and discoverer of oxygen. However, he would eventually fall foul of the increasing attention that chemists attracted.

Putting the Fizz into Water

The discoveries of Joseph Black and Henry Cavendish signaled exciting times for pneumatic chemistry, and no one was more prolific than Joseph Priestley (1733–1804), discovering no fewer than eight new gases. Working as a minister and teacher, Priestley came from a background of dissenters—religious radicals who "dissented" from mainstream Anglicanism and often had radical political ideas.

After meeting the American scientist Benjamin Franklin in 1766 he pursued scientific research, and shortly after got a job as a minister in Leeds, England, living next to a brewery. He showed that the layer of air bubbling out of vats of fermenting beer was "fixed air" (carbon dioxide). And as he had a large supply at his disposal, he decided to try and simulate the natural fizz of some mineral waters. Dissolving the carbon dioxide under pressure in water, he created carbonated water, sparking a European craze for "soda water."

Gas Man

In 1773 Priestley was offered a position that gave him plenty of time for research. He delved further into pneumatic chemistry, perfecting Hale's pneumatic trough (see p. 77) and cleverly using mercury instead of water to collect gases that were water-soluble. He also acquired a magnifying glass 1ft (30cm) across, which he used to focus the sun's rays and thereby generate very high temperatures. Using this apparatus he discovered gases including nitrogen monoxide (NO), dinitrogen monoxide (N_2O), also known as laughing gas, sulfur dioxide (SO_2), and ammonia (NH_3).

By 1772 Priestley had arguably made the first observation of photosynthesis, showing that plants produced an "air" that animals needed for respiration. In 1774 he synthesized this air using his glass to heat red calx of mercury, a powder produced by burning mercury in air. By making the calx hot enough, he turned it back into mercury. He noticed the gas given off was colorless and odorless but caused a flame to burn very brightly. Further testing revealed that it was "superior" to common air: a mouse lived for an hour in a glass vessel containing the air; but if it had been common air, Priestley explained, the mouse would have lived for about

quarter of an hour. Priestley was committed to the phlogiston theory, so identified his new gas as dephlogisticated air. But on a trip to Paris in 1774 he told Lavoisier about his findings (see pp. 86–87), which led to Priestley's "superior air" being identified as oxygen, and the dismantling of phlogiston doctrine.

A TARGET OF HIS BELIEFS

Science became highly political in the 18th century, and few scientists were more political than Priestley. As an outspoken opponent of the established Church and supporter of the French Revolution, he became a target of anti-Revolutionary feeling in England. On July 14th, 1791, the second anniversary of the Revolution, rioters in Birmingham burned down Priestley's home. He fled to London with his family, and was eventually forced into exile, moving to America. He ended his days in isolation, insisting on clinging to the phlogiston theory, which put him at odds with scientists in the chemistry mainstream.

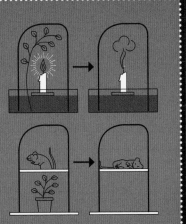

Priestley carried out experiments that showed that plants produce oxygen. He discovered that a burning candle and a mouse could only survive in a sealed jar of air if a plant was also placed in the jar. Both the candle and the mouse relied on the plant to produce oxygen. (And in turn the plant relied on the carbon dioxide they produced.)

Lavoisier and the Chemical Revolution

The greatest chemist of his age—perhaps ever—discovered no new elements, yet did more than anyone to finally transform chemistry into a science. Known as the father of the "chemical revolution," Antoine-Laurent Lavoisier worked out the role of oxygen, defined the term "element" and introduced scientific nomenclature, before his life was cut short by the French Revolution.

Tools for the Trade

Antoine-Laurent Lavoisier (1743–1794) was the son of a rich lawyer, expensively educated and trained in law before taking up science,

Lavoisier's ability to buy **the best equipment** helped him to become a superior chemist.

initially as a geologist. This led him to chemistry and he set up a laboratory with the aim of entering the prestigious Academy of Sciences, which he succeeded in. To secure a private income that would fund his researches, Lavoisier fatefully joined the Ferme Générale, a private consortium that collected taxes for the crown (see box, p. 89). Science was becoming increasingly specialist and therefore expensive—especially the precision instruments that Lavoisier would need.

In 1772 he turned his attention to pneumatic chemistry, experimenting with combustion of phosphorus and sulfur, which, he discovered, gained weight when burned in air. He also found that when he heated litharge (lead[II] oxide, an ore of lead) with charcoal (carbon), it reduced to lead, released gas, and lost weight.

This finding was at odds with the phlogiston theory, which claimed that the reduction of ore to lead

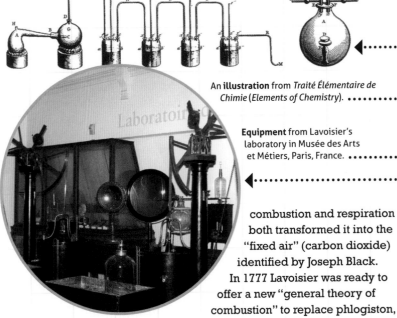

An **illustration** from *Traité Élémentaire de Chimie* (*Elements of Chemistry*).

Equipment from Lavoisier's laboratory in Musée des Arts et Métiers, Paris, France.

combustion and respiration both transformed it into the "fixed air" (carbon dioxide) identified by Joseph Black.

In 1777 Lavoisier was ready to offer a new "general theory of combustion" to replace phlogiston, along with a new name for his new principle of combustion: oxygen. His research with acids had shown that oxygen was present, leading him to call his "eminently respirable air" the "acidifying principle" or "oxygen principle." (Oxygen derives from the Greek for "acid-maker.")

involved addition, not subtraction, and it set Lavoisier on the path to destroying the myth of phlogiston.

The Secret of Combustion

Two years later, Lavoisier learned of Priestley's discovery of "dephlogisticated air." Experimenting with the new gas for himself, he soon realized that this was the common principle underlying the processes of combustion, reduction, respiration, fermentation, and acidity. As Priestley had also done, Lavoisier proved that the new air formed that portion of the atmosphere that supports life, which led him initially to term it "eminently respirable air," and he was able to show that

Armed with this new concept of oxygen, Lavoisier was able to show that the phlogiston theory was back to front. Combustion, respiration, and rusting involved the addition of oxygen and reduction its loss. Fixed air was produced by combining charcoal with oxygen. When he learned how to make water by burning hydrogen in oxygen, the final piece of the puzzle fell into place, and Lavoisier was able to show that water was a compound, combining hydrogen (which Lavoisier named from the Greek for "water-maker") with oxygen.

Scientific Definitions

The culmination of Lavoisier's chemical career was the publication of his *Traité Élémentaire de Chimie* (*Elements of Chemistry*) in 1789. It set out his findings and his reasoning in clear, logical style, supporting his modern, scientific vision of chemistry: "We must trust to nothing but facts: These are presented to us by Nature, and cannot deceive. We ought, in every instance, to submit our reasoning to the test of experiment, and never to search for truth but by the natural road of experiment and observation."

Among the important innovations of Lavoisier's chemistry was a new and conclusive definition of an element: "the last point which analysis is capable of reaching"—in other words, those substances that couldn't be broken down any further. He admitted that as technology advanced, some substances previously incapable of decomposition might be shown to be compounds, and indeed several of the substances on his list of 33 elements proved to be oxides. He also predicted that several alkaline earths (basic solids that couldn't be broken down any further) would prove to be metal oxides, and sure enough Humphry Davy later used the new technology of electrolysis to isolate the alkaline earth metals from their molten salts (see pp. 112–113).

Lavoisier's "balance sheet" approach was another great contribution to chemistry as a science. Using his expensive, highly sensitive instruments, he perfected the art of measuring both reactants and products, whether solid, gas, or liquid, and stressed the importance of quantifying exactly what went into and came out of a reaction. This led him to define the principle of conservation of matter (see box).

"We may lay it down as an incontestable axiom, that, in all the operations of art and nature, nothing is created; an equal quantity of matter exists both before and after the experiment; the quality and quantity of the elements remain precisely the same; and nothing takes place beyond changes and modifications in the combination of these elements. Upon this principle the whole art of performing chemical experiments depends. We must always suppose an exact equality between the elements of the body examined and those of the products of its analysis."

—Antoine-Laurent Lavoisier

LAVOISIER'S BIGGEST MISTAKE

Lavoisier wasn't incapable of making mistakes. A central aspect of his new system was a hypothetical principle of heat, which he called caloric. Though an undetectable substance without mass, caloric was supposed to act like a liquid or gas, and Lavoisier claimed that oxygen gas was actually a compound of oxygen and caloric, with caloric accounting for its phase. This was as much a dead-end as phlogiston, and in many ways simply a renamed version of the classical element of fire (see pp. 82-83).

REVENGE OF THE NERD

Lavoisier was very accomplished. As well as his chemical research, he was a diligent tax collector and got involved with many civic and government issues, including helping devise the metric system and assisting the Parisian authorities with building an unpopular antismuggler wall around the city. Unfortunately none of his accomplishments rescued him in the face of a vengeful amateur. Before becoming a radical leader of the French Revolution, Jean-Paul Marat (1743-1793) was an aspiring scientific dabbler with hopes of joining the Academy of Sciences, but Lavoisier had blocked his membership. Now a man of terrible power, Marat accused the chemist of attempting to "imprison" Paris with his wall. At his trial, the judges dismissed Lavoisier's pleas for mercy so he could pursue his research, and his support for revolutionary causes counted for nothing when set against his membership of the hated tax-farming operation. He was executed by guillotine on May 8th, 1794.

Oxygen and Redox Reactions

The oxygen concept introduced by Lavoisier changed chemistry and has proved to be one of the most important principles in the field. Not only is oxygen a vitally important element, but the process by which it forms bonds and ions—the reduction-oxidation or "redox" reaction—is central to the chemistry of combustion, electrochemistry, respiration, photosynthesis, and acids and bases.

Oxygen Rules

We now know that oxygen has six electrons in its outer shell and so needs two in order to fulfill the octet rule—in other words it has a valency of 2 (see pp. 64–65). This means that when other elements bind with oxygen (oxidation), they do so by donating two electrons, and when a compound loses oxygen (reduction) it gets those electrons back.

reduced compound A   oxidized compound B

A

B

A is oxidized, losing electrons

B is reduced, gaining electrons

A

B

oxidized compound A   reduced compound B

Oxidation is matched by reduction.

However, reduction and oxidation now mean more than just losing or combining with oxygen: they refer to any reactions where electrons are gained or lost. Since you can't have one without the other, reduction and oxidation are two sides of the same coin—half-reactions that together make up the redox reaction. The most important rule to remember is: reduction = gain of electrons, and oxidation = loss of electrons.

Extended Definitions

Redox reactions include ones we've already met, such as combustion and rusting, neutralization, displacement reactions, and electrochemical reactions. The wide range of processes involving redox reactions means that reduction and oxidation can be defined in three ways:

Reduction can mean gain of electrons, but it can also mean loss of oxygen or gain of hydrogen—all three are equivalent, because they involve a net gain of negative charge. So when a zinc cation (positively charged ion) becomes zinc metal, it has been reduced through gain of electrons. When red mercury calx (mercury oxide, HgO) is heated until it decomposes into mercury and oxygen, it has been reduced through loss of oxygen. And when carbon monoxide (CO) and hydrogen gas (H_2) are combined to produce methyl alcohol (CH_3OH), the carbon monoxide has been reduced through gaining hydrogen.

Oxidation can mean loss of electrons, gain of oxygen, or loss of hydrogen. So when sodium and chloride are combined to form table salt ($Na + Cl \rightarrow NaCl$), the sodium has been oxidized by losing an electron to the chlorine. When carbon burns, it is oxidized to carbon dioxide by gaining oxygen atoms. And when the methyl alcohol reaction is reversed ($CH_3OH \rightarrow CO + 2H_2$), methyl alcohol has been oxidized to carbon monoxide by losing hydrogen.

Displacement Reactions

In all these reactions, the reduction of one substance has been matched by the oxidization of the other and vice versa. A good illustration of this is a displacement reaction, such as copper displacing silver from a silver nitrate solution:

$$Cu(s) + 2AgNO_3(aq) \rightarrow$$
$$Cu(NO_3)_2(aq) + 2Ag(s)$$

What actually happens in such a reaction is that the silver nitrate is split into ions ($Ag^+ + NO^-_3$) because it's in solution, and the nitrate ions don't take part in the reaction. The silver ion oxidizes the copper (known as an oxidizing agent), while the copper reduces the silver. This equation shows only the active ions:

$$Cu(s) + 2Ag^+(aq) \rightarrow$$
$$Cu^{2+}(aq) + 2Ag(s)$$

Breaking this down still further to its half-reactions makes clear the electron transfer involved in the redox reaction (electrons are e^-):

$$Cu(s) \rightarrow Cu^{2+}(aq) + 2e^- \text{ [oxidation]}$$
$$2Ag+(aq) + 2e^- \rightarrow 2Ag(s) \text{ [reduction]}$$

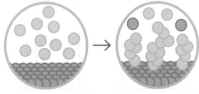

Copper (Cu) displacing silver (Ag) from a **silver nitrate** solution.

Hydrogen and "Ballomania"

Hydrogen is in many ways the primal element—the first in the periodic table as it was the first created element in nature, produced by the Big Bang. It remains the most common element in the universe, making up the vast majority of the cosmos. Here on Earth it could hold the key to a greener future, though its best-known application has been lighter-than-air flight.

Vital Element

Hydrogen makes up roughly three-quarters of the mass of the universe and more than 90% of all molecules. Although first described as a new element by Cavendish, it had previously been produced by medieval alchemists, as they commonly worked with strong acids and metals—a combination that produces hydrogen. In the 17th century, Frenchmen Theodore Turquet de Mayerne (1573–1655) and Nicolas Lemery (1645–1715) produced hydrogen by adding iron to sulfuric acid and noted it was highly flammable, but assumed it

The **first recorded human balloon flight** took place in Paris in 1783 in a giant hot air balloon designed by the brothers Joseph and Etienne Montgolfier (pictured left).

was a form of sulfur. Not until Lavoisier decomposed water was hydrogen understood to be an element and given its name.

Cavendish had blown soap bubbles with hydrogen and observed their buoyancy, and Lavoisier measured this precisely, finding that hydrogen weighed only $\frac{1}{13}$ as much as common air. It didn't take long for the wider community to see an obvious use for the new gas. In 1782 Joseph-Michel Montgolfier (1740–1810), who owned a papermaking company with his brother, considered the idea of using gas-filled paper bags for military air assaults. The following year the Montgolfier brothers used hot air as their buoyancy aid, but that August in Paris, the scientist Jacques Alexandre Charles (1746–1823) achieved a much faster and higher ascent with a silk bag filled with hydrogen.

FUEL OF OUR FUTURE?

Hydrogen may well prove to be the energy of the future. Hydrogen can be produced through decomposition of water. And in turn, hydrogen and oxygen can be combined in a fuel cell to produce electrical energy. Its only product is water, so it's pollution-free. The idea of hydrogen as fuel isn't a recent one. As early as 1874, a character in Jules Verne's novel announced: "I believe that water will one day be employed as fuel, that hydrogen and oxygen which constitute it, either singly or together, will furnish an inexhaustible source of heat and light... Water will be the coal of the future."

The Balloon Race

In September the Montgolfier brothers caused a sensation at Versailles by launching a hot air balloon carrying a sheep, a duck, and a cockerel. France, with the rest of the world not far behind, was caught in the grip of "ballomania." The world waited to see who would be the first to send up a man.

In Paris on November 21st, the first manned ascent took place in a "Montgolfier," as hot air balloons were now known. Just 10 days later Charles took to the skies in a hydrogen balloon that had many features of a modern balloon, including a wicker basket, an airtight balloon coated with rubber, a gas-venting system, and a ballast system. His ascent attracted a crowd of 400,000 people—half of Paris. Despite all the excitement, there were doubts about the usefulness of the new contraptions. But Joseph Gay-Lussac made a notable ascent in 1804, reaching 23,000ft (7km) above Paris in a hydrogen balloon and thereby discovering the limit at which air was still breathable. Ballomania tapered off but hydrogen was still used for balloons up until the Zeppelin era in the 1900s, after which it was replaced by helium.

A Scientific Naming System

Among the legacies of Lavoisier and his French school was the introduction of a new language for chemistry—a scientific language. By ditching the old terminology and naming system that had built up haphazardly during the evolution of chemistry, the new system brought clarity of definition and clarity of thought that are still with us today.

Chemical Chaos

Over the millennia, alchemy and industry had built up a colorful and chaotic naming system. Names came from different traditions, languages, and regions. They could derive from locale, method of manufacture, subjective properties (such as taste, smell, consistency, and color), the person who had discovered them, or factors such as astrology or magical influence. A single chemical could have multiple names, reflecting different historical sources or simply different production methods. Nitric acid, for instance, was known as spirit of niter when distilled from saltpeter, or as *aqua fortis*. Terms such as "earth," "oil," and "air" weren't specific or consistent, and names might change for the same substance in different phases or in solution.

An Improved Language

In the 18th century, the discovery of new elements and compounds focused attention on the problem, and attempts were made to reform and standardize the nomenclature. Swedish chemist Torbern Bergman came up with a similar system to botanical nomenclature, and this influenced the French chemist Louis-Bernard Guyton de Morveau (1737–1816). He proposed that chemical names should be short, based on classical roots, and reflect the composition of a substance.

Louis-Bernard Guyton de Morveau (1737–1816).

Guyton's ideas were realized in 1787 with the publication of the *Method of Chemical Nomenclature*, by himself, Lavoisier, and two fellow chemists. However, the system set out in the book was controversial because it was based on Lavoisier's theories—theories seen as unproven by many chemists, especially in Britain and Germany. For instance, the names of mixed bodies (compounds) would be created by combining the names of the simple bodies (elements) they were composed of. But this depended on Lavoisier's definition of an element and his claim that well-known substances such as water were actually mixed bodies.

Under the new scheme, litharge or white lead became lead oxide, while stinking gas became sulfuretted hydrogen gas. There was no place for phlogiston, and only 33 substances were included as elements. Recently and newly discovered elements, such as oxygen and hydrogen, were named for their chemistry rather than subjective properties, but again, these names reflected Lavoisier's theories. Word endings showed proportions—so, for instance, sulfur*ic* acid contained more of the acid principle (oxygen) than sulfur*ous* acid.

For Lavoisier, the reformation of nomenclature was essential for a scientific chemistry: "a well-made science depends on a well-made language."

NEGATIVE RESPONSE

The new system's dependence on Lavoisier's theories meant it received a rough ride when it was introduced. But German and British chemists were forced to learn its principles in order to read Lavoisier's books, and the system caught on. It even survived revisions of the Frenchman's theories—for instance, when Humphry Davy proved that hydrochloric acid contains no oxygen, so demolishing Lavoisier's theory of oxygen as the acid principle, the name should have been changed. The British accepted the new classical names, but in Germany many of them were translated, so that in German, oxygen is still *Sauerstoff* ("acid stuff") and hydrogen *Wasserstoff* ("water stuff").

"As ideas are preserved and communicated by means of words, it necessarily follows that we cannot improve the language of any science, without at the same time improving the science itself; neither can we…improve a science without improving the language or nomenclature which belongs to it."

—Antoine-Laurent Lavoisier,
Elements of Chemistry

4

Atoms and Ions

Lavoisier's chemical revolution changed the way chemistry was done and inspired a new generation of scientists. Yet the young science still lacked many of the attributes Newton had brought to physics, such as simple mathematical principles and laws that would make it a truly quantitative science. This chapter explains these principles and laws, and tells how they came to be discovered, introducing along the way an electrifying new tool for the analysis of matter.

Atomic Weights and Atomic Theory

Although chemistry was now firmly established as a science, many aspects remained mysterious. Which substances were elements? What were elements made of? How did they come together to make compounds and how could chemists work out the formulae of those compounds? The microcosmic world of matter seemed impenetrable, until a simple discovery brought it all into focus.

Definite Proportions

Robert Boyle and others had revived atomism in the 17th century, and most of the scientific world came to accept Isaac Newton's speculations. Newton imagined matter to be separate, indivisible particles, which interacted via forces of attraction and repulsion similar to gravity but on a much smaller scale. In other words, the microcosmic world of atoms mirrored the macrocosmic world of planets and moons.

This seemed a plausible theory but without the powerful technology needed to explore the atomic world directly, it was of little practical use until the advent of chemical atomism. In 1788 French chemist Joseph-Louis Proust (1754–1826) discovered the "law of constant composition," also known as the "law of definite proportions." Previously it had been accepted that the composition of a compound could vary—so that, for instance, some bodies of water might have more oxygen than others. Proust's careful

> "Chemical compounds contain the same proportions by mass, regardless of the source or the amount."

Joseph-Louis Proust (1754–1826) famously discovered the "law of constant composition."

analysis showed that this wasn't the case. Compounds all consisted of elements in definite simple ratios by weight, and what was more, the proportions were whole numbers.

A New System

John Dalton (see pp. 100–101) realized that Proust's law could be explained by atomic theory: compounds must be formed through the combination of separate particles, and these particles must vary in weight by multiples of whole numbers. In 1808 Dalton published *A New System of Chemical Philosophy*, the founding text of chemical atomism. In it he explained that each element has its own characteristic atoms, distinguished by their relative weights. He didn't speculate on other properties of atoms—there was no way to examine these—but using careful quantitative chemistry he worked out the relative weights of atoms of different elements.

Hydrogen was the lightest element known, so Dalton gave it an atomic weight of 1, and calculated other elements accordingly. Analysis of water had shown that its components were oxygen and hydrogen in a ratio by weight of 8:1. Dalton assumed that nature kept things as simple as possible, so the most likely formula for water was the simplest—one atom of hydrogen to one atom of oxygen. Accordingly, oxygen must have a relative or atomic weight of 8. Using these weights, he could work out the weights of other elements. In practice Dalton's assumptions were

MASS VS. WEIGHT

The terms "atomic mass" and "atomic weight" are sometimes used interchangeably. But although both are measured in atomic mass units (amu), there is a difference.

Atomic mass is the mass number of an atom—the total of protons and neutrons. Atoms of the same element with different numbers of neutrons, and hence different mass numbers, are called isotopes (see pp. 132-133). In nature, elements generally exist as a mix of isotopes. For example, carbon exists as three different isotopes: carbon-12, carbon-13, and carbon-14, with atomic masses of 12, 13, and 14.

Atomic weight is the *average* of the isotopes' atomic masses, depending on which isotope is the most abundant. Carbon mostly exists as carbon-12, so the average atomic mass of carbon is 12.011 (12), and this is the atomic weight.

atomic mass

$$^{12}_{6}C$$

atomic weight

often wrong and this threw off his calculations, so few of his atomic weights proved accurate, but he had established chemical atomism and set chemists on the road to quantifying their science.

John Dalton

A humble man from Britain's provinces, John Dalton had little formal education and was isolated from the centers of intellectual power, yet he became famous despite himself. His discoveries helped advance the development of chemistry, while his career marked an important stage in the evolution of science in general.

Small-town Scientist

Born into a family of devout Quakers in Cumbria in rural northern England, John Dalton (1766–1844) was destined to be an outsider in the world of 19th-century British science—a world of privilege, gentleman-amateur scientists, and an establishment firmly centered on the Royal Society in London.

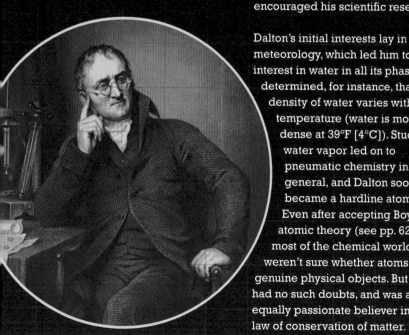

As a Quaker, a member of a non-established Church, Dalton was barred from attending the great universities even if he could have afforded it. Instead his education was limited to the village school, which he attended until he was 12 and then started teaching there. Later he became a headmaster, a lecturer, and finally a private tutor, moving to Manchester, where fellow Quakers encouraged his scientific research.

Dalton's initial interests lay in meteorology, which led him to an interest in water in all its phases. He determined, for instance, that the density of water varies with temperature (water is most dense at 39°F [4°C]). Studying water vapor led on to pneumatic chemistry in general, and Dalton soon became a hardline atomist. Even after accepting Boyle's atomic theory (see pp. 62–63), most of the chemical world weren't sure whether atoms were genuine physical objects. But Dalton had no such doubts, and was an equally passionate believer in the law of conservation of matter.

Dalton's Law

This atomist view of gases led him to formulate what is now known as Dalton's law, or "Dalton's law of partial pressures," which he published in 1801. It states that in a mixture of gases, each individual gas exerts pressure independently of each other. So the total pressure is the sum of the partial pressures of the individual gases. This only holds true for ideal gases (see pp. 56–57) and so more accurately the law was that in gas mixtures, such as the atmosphere, the individual gases have no chemical interaction with one another.

Dalton's theory of atomic weights brought greater fame, but as an outsider from the provinces who was reluctant to join the Royal Society, which he considered a group of amateurs, he wasn't always well thought of by contemporaries. Despite this, Dalton gained international renown and 40,000 people attended his funeral.

A NEW BREED OF SCIENTIST

Dalton was one of a new breed; men of no status or family wealth who became professional scientists, at a time of growing tension between the provinces and the gentleman-amateur scientific establishment based in London. He played an important role in setting up the British Association for the Advancement of Science (known as BA), which provided an alternative to the Royal Society and reflected the evolving professionalism of the field. Established in 1831, it met annually, mostly outside London, and was the forum where major advances in British science were announced.

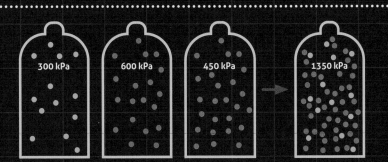

The **total pressure** in kPa (kilopascals) of a mixture of ideal gases is equal to the sum of the partial pressures of the component gases.

Moles and Avogadro's Number

Atoms and molecules are extremely small and weighing or counting them directly isn't possible. This is where the mole comes in, giving chemists a way of relating atomic weights to real-world weights and measures, and working out the actual formulae of compounds. The mole is the bridge between the microscopic and macroscopic worlds.

Counting by Weighing

A mole is an amount of a substance that contains as many particles as there are atoms in exactly 12g of the isotope carbon-12. The particles can be anything—atoms, molecules, ions, electrons—but they must be specified. The number of atoms in 12g of carbon-12 is known as the Avogadro number, named for the 19th-century scientist (see pp. 104–105) who originated the concept of the mole. It is 6.0221367×10^{23}, which is 602 billion trillion, or 602 with 21 zeros after it. Moles allow us to count by weighing. The mole gives the atomic weight of an element in grams—quantities that chemists can work with in the real world. The mole

isn't simply atomic weight expressed in grams; it can also be the molecular, or formula, weight expressed in grams. "Formula weight" is the sum of the atomic weights of atoms in a compound. For instance, the atomic masses of the atoms that make up a molecule of water (H_2O) are 1 atomic mass unit (amu) each for the 2 hydrogen atoms and 16 amu for the oxygen, giving a total of 18 amu. So the formula mass of water is 18 amu. A mole of water therefore weighs 18g. (The atomic weights of these elements are not quite round numbers because of isotopes, but have been rounded up here for simplicity.)

6.0221367×10^{23}

The Unknown Number: Avogadro had no way of working out the number that now bears his name. The first man to name it was the French physicist Jean Perrin (1870–1942) in 1908.

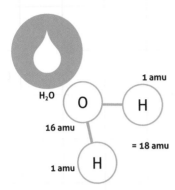

H₂O

1 amu

O

16 amu

H

= 18 amu

1 amu H

The **molecular structure of H₂O**.
The total atomic mass of a water
molecule is 18 amu.

Moles in Action

The mole is a powerful tool for
chemists. Say you have 22.99g of
sodium (Na) and you want to make
table salt by combining it with
chlorine (Cl), but you don't want to
waste reactants by having too much
of one or the other. How do you know
how much chlorine to use? Handily
the atomic weight of sodium is 22.99
amu, so you know you've got 1 mole
of sodium. Given that you also know
the formula of table salt is NaCl, you
know that you need one atom of
chlorine for each atom of sodium—
or 1 mole of chlorine for each mole
of sodium. The atomic weight of
chlorine is 35.453, therefore you
need 35.453g of chlorine. In practice,
no reaction is 100% efficient so not
every particle of reactant will react,
but you get the idea. Weighing gases
is time-consuming, but helpfully it's
possible to convert the mole concept
into measures of liquid and gaseous
volume (see box).

MORE MOLE MEASURES

Terms related to the mole concept
are molar mass and molar volume:

Molar mass is the mass of a mole of
that substance, expressed as grams
per mole (g mol⁻¹).

Molar volume (V_M) is the volume
occupied by one mole of that
substance, which depends on its
density. This in turn depends on
temperature and pressure, though
the density of liquids varies little, so
molar volumes for liquids at room
temperature and sea level are useful
across a wide range. For gases, V_M
depends on the precise temperature
and pressure. At standard
temperature and pressure, the V_M
for any gas is 22.4 dm³.

Moles allow us to count by
weighing. The mole gives
the atomic weight of an
element in grams—
quantities that chemists can
work with in the real world.

Amedeo Avogadro

The man who made the vital link between the microscopic and macroscopic worlds with a daring new concept was ignored and dismissed in his own time. Despite training as a lawyer, Amedeo Avogadro took private lessons in math, chemistry, and physics before starting a career in science.

A Leading Light

John Dalton's work with atomic weights had run into a seemingly insurmountable obstacle. Although it was possible to find out relative proportions of elements in a compound, there was no way to relate this to the formula of that compound. For example, Dalton had assumed that water was a 1:1 mix of hydrogen and oxygen, but he had got the atomic weight of oxygen wrong and this undermined his whole system.

It took Amedeo Avogadro (1776–1856), a modest nobleman from northern Italy, to show the way forward, leading to the concept of the mole, which made it possible to calculate actual atomic weights and thereby empirical formulae (see pp. 110–111). Avogadro practiced law until 1800, when he began his scientific education, becoming a mathematical physicist.

Bridging the Gaps

Avogadro's breakthroughs were based on two discoveries by the French chemist Joseph Gay-Lussac (1778–1850). The first was that all gases expand equally as temperature increases. The second was the "law of combining volumes," which said that the volumes of gases that react with one another are ratios of small whole numbers. For example, 2 volumes of hydrogen + 1 volume of oxygen → 2 volumes steam. Although Dalton didn't realize it, Gay-Lussac's law was the equivalent of Proust's "law of constant composition" (see pp. 98–99) and a confirmation of his atomic theory.

> "Equal volumes of all gases, under the same temperature and pressure, contain the same number of smallest particles."
>
> —**Amedeo Avogadro**

However, Avogadro did make this connection. He boldly stated that Gay-Lussac's first discovery meant that "equal volumes of all gases, under the same conditions of temperature and pressure, contain the same number of smallest particles"—now known as Avogadro's law. He would later call these particles "molecules" and suggested that gases such as oxygen and hydrogen might be molecules with two atoms. Using his law and the law of combining volumes, Avogadro worked out that hydrogen and oxygen must combine to form water in the ratio 2:1, and therefore the molecular formula of water must be H_2O. This in turn made it possible to calculate the correct atomic weights of the elements.

Unfortunately Avogadro's brilliant idea made no impact at the time (see box). Not until 1860, when he was already dead, did the Italian chemist Stanislao Cannizzaro (1826–1910) demonstrate the power of Avogadro's hypothesis and force the scientific community to rethink. Instead, debate and confusion over atomic weights and molecular formulae continued

AN UNSUNG HERO

So why were Avogadro's ideas ignored? It was probably a combination of factors. Avogadro's base of Turin was far from the centers of scientific power at the time. He also had a reputation as a poor experimentalist, which meant that other chemists didn't take him seriously, and he damaged his own cause by failing to support his hypotheses with hard data. Even his claim that oxygen and hydrogen were diatomic went against him, because the dominant theory of the day was that of Jöns Berzelius (see pp. 108-109), which stated that atoms of the same element would repel each other.

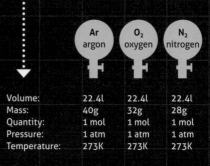

	Ar argon	O_2 oxygen	N_2 nitrogen
Volume:	22.4l	22.4l	22.4l
Mass:	40g	32g	28g
Quantity:	1 mol	1 mol	1 mol
Pressure:	1 atm	1 atm	1 atm
Temperature:	273K	273K	273K

Avogadro's law: mass and molecular formula of a gas may vary, but the volume occupied by 1 mole at standard pressure and temperature is a constant at 22.4 liters/mol

Ions and Charges

The advent of electrochemistry opened up an exciting new field of science. To understand this pioneering technology, let's take a quick review of the basics of ions and charges, including the names used for different types of ion—they may appear complicated, but in fact they're quite logical.

Finding Stability

As we saw in Chapter 2 (see pp. 64–65), atoms achieve a stable electronic configuration by losing or gaining electrons from their outer valence shells. When an atom donates or receives one or more electrons, there will be a mismatch between the number of protons it has and the number of electrons. This atom gains a positive or negative charge, and it becomes known as an ion.

Atoms that form ions lose or gain electrons following the octet rule— they seek to have the same electron configuration as the noble gas that is nearest in the periodic table. For instance, when table salt (NaCl) is formed, sodium donates an electron to chlorine. The sodium atom becomes a cation (positively charged ion) with a charge of +1, matching the electronic arrangement of neon, and the chlorine atom becomes an anion (negatively charged ion) with a charge of -1, matching the electronic grouping of

argon. Charges are shown by a superscript after the element name—so salt contains Na^+ and Cl^- ions.

Ionic compounds form through electrostatic attraction between positively and negatively charged ions, creating an ionic bond. A typical example of an ionic compound is a salt, formed when an acid reacts with a base, usually a metal. Metal salts generally form crystal lattice structures.

Diverse Species

"Species" is a term chemists use to describe types of ion, which include "monatomic" and "polyatomic" ions. The species of ion that an element or compound can produce is governed by periodic law (see pp. 124–125). Among monatomic ions, for instance, alkali metals form cations with a charge of +1, while oxygen and sulfur form anions with a charge of -2. Anion names end in "-ide," so the anions of oxygen and sulfur are *oxide* and *sulfide* ions.

The transition metals can have different oxidation states (see pp. 90–91). This means they can form ions with different positive charges, where the charge is equal to the oxidation state. The oxidation state (and therefore charge) is usually shown by Roman numerals in brackets, but it can also be shown using naming conventions. The name for the ion with the lower oxidation state name ends in "-ous." So: iron(II), Fe^{2+}, is *ferrous*, while iron(III), Fe^{3+}, is *ferric*.

There are many polyatomic species, most of them "oxyanions"—anions containing oxygen. Oxyanion names commonly end in "-ate." Those that have the same charge but fewer oxygen atoms end in "-ite." So, SO_4^{2-} is *sulfate*, but SO_3^{2-} is *sulfite*. Other important polyatomic anions include hydrogen carbonate, also known as bicarbonate (HCO_3^-), nitrate (NO_3^-) and nitrite (NO_2^-), hydroxide (OH^-), cyanide (CN^-), and peroxide (O_2^-).

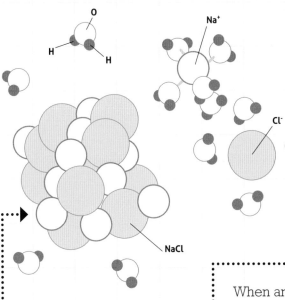

Table salt (NaCl) is a crystal held together by strong ionic bonds. When we dissolve salt by stirring it in water, the ionic bonds break and the ions are released into the water. Each Na^+ and Cl^- ion attracts a shell of water molecules that prevent the ions from reforming into a crystal. This process is called "hydration."

When an atom donates or receives one or more electrons, there will be a mismatch between the number of protons it has and the number of electrons. This atom gains a positive or negative charge.

Jöns Jacob Berzelius

After Lavoisier the next great figure in chemistry was Jöns Jacob Berzelius, who made many discoveries and advanced the theory and practice of his science. After exploring the new art of electrolysis, he went on to perfect techniques of quantitative chemistry, discover new elements and compounds, devise a system of notation, and effectively control chemistry in Europe.

In 1803 Berzelius stuck electrodes into a solution of a neutral salt and noticed that its "acid" component formed around the positive pole and its "base" around the negative pole. A few years later, Humphry Davy used electrolysis to isolate sodium, potassium, and the alkaline earths (see pp. 114–115). This convinced Berzelius of the importance of electricity in decomposing compounds and binding elements. On the basis of this, he worked out a "dualistic theory," classifying all substances as either electropositive or electronegative. Salts, he claimed, were formed when electronegative substances acted as acids and electropositive ones acted as bases. He was also convinced that oxygen was an essential constituent of all acids and bases.

The Power of the Pile

Born in Väversunda in Sweden, Jöns Jacob Berzelius (1779–1848) overcame early difficulties in accessing education by reading every chemistry textbook he could lay his hands on. Although he trained as a doctor, his true passion was chemistry, especially the new field of electrochemistry, which had been sparked into life by Volta's 1800 invention, the voltaic pile (see box).

> Berzelius dramatically improved the calculation of atomic weights and molecular formulae.

Precision and Influence

Berzelius dramatically improved the calculation of atomic weights and molecular formulae by achieving new standards of accuracy in quantitative chemistry. And he would go on to prepare, purify, and analyze more than 2,000 substances, including several new elements. However, his dualistic theory led him to reject Avogadro's theories about diatomic molecules, which resulted in much confusion over the atomic weights and formulae of some important elements, especially gases.

Among other achievements, Berzelius established that organic compounds follow the same rules of proportional composition as inorganic ones, and helped characterize some important phenomena in organic chemistry. But perhaps his most remarkable achievement was the way in which he managed to dominate European chemistry from about 1820, even though his Stockholm base was far from the centers of European science. Through writing a constantly updated standard textbook, and editing an essential yearbook, Berzelius became the gatekeeper for chemistry. In later years, though, he became increasingly set in his ways, obstructive, and bitter at being sidelined.

THE VOLTAIC PILE

The invention that revolutionized many of the sciences was fairly simple—a stack of alternating silver and zinc disks interleaved with cards soaked in brine. This primitive battery, named the voltaic pile after its Italian creator Alessandro Volta (1745-1827), was able to generate enough voltage for electrolysis. When they learned of the new device, physicist William Nicholson and surgeon Anthony Carlisle immediately made their own, using it to decompose water into oxygen and hydrogen. The two Englishmen printed the results before Volta's paper about his device had even been published.

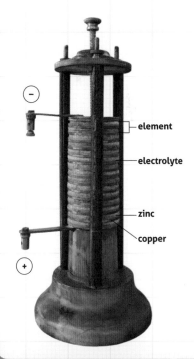

— element

— electrolyte

— zinc

— copper

Chemical Notation

The dream of 19th-century chemists was to bring the same rigorous precision as mathematics to their science, in the same way that Newton had transformed physics. Any modern student of chemistry knows that they succeeded, as nothing is now more symbolic of this achievement than the chemical equation. Bridging this gap was one of Berzelius's lasting achievements.

The Need for Signs

In 1813 Berzelius began to put together a new system of notation that used signs to express chemical proportions. As with nomenclature (see pp. 94–95), previous notation methods had been haphazard, reflecting the uneven development of chemistry over thousands of years and across many cultures and languages. Alchemists had used symbols dense with meaning, often from mystical sources, but the discovery of new substances and the new understanding of elements called for a new approach.

John Dalton had put together his own system with a series of simple diagrams, but it had obvious flaws. Berzelius explained that chemical signs should be letters, and he decided to use the first letter of an element's Latin name. For example, sulfur = S. If an element had the same initial letter as another, he used the first two letters: silicon = Si. And if these were common to two elements, he used the first letter and the first consonant they didn't have in common. For example: stibium (antimony) = St and stannum (tin) = Sn. The periodic table on pages 118–119 shows the one- or two-letter combinations for every element. This system appealed to printers because they could use letters they already had, and it soon became the universal standard.

An engraving of various **alchemical symbols**.

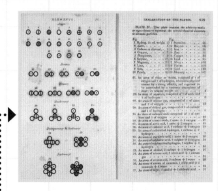

Dalton devised a new **system of notation** for the elements but it was unwieldy and would have required printers to make new letters for the symbols, so it never caught on.

represent one volume, or mass, of a substance, with multiples represented by coefficients (numbers in front of the symbols). Now chemical equations could be written like mathematical ones. In an equation the two sides are separated by an arrow →, which shows the direction from reactants to products. Many chemical reactions are reversible: they can go both ways, even though one direction may happen faster than the other. Eventually the two will occur at equal rates. In such cases a two-way arrow is used: ⇌

In the modern version of chemical notation, subscripts *after* the element symbol show the number of atoms in the molecule, while superscripts represent positive or negative charge, in the case of ions. Subscripts *before* the element symbol represent atomic number and superscripts represent mass. The atomic mass also indicates the isotope of the element—so 12C, for instance, is carbon-12.

The "law of conservation of matter" has an important consequence for chemical equations: because atoms cannot be created or destroyed, there must be the same number of atoms on one side of the equation as on the other. In other words, chemical equations must balance. For instance, take the equation for hydrogen + oxygen = water. As both hydrogen and oxygen are diatomic, you might write:

$$H_2 + O_2 \rightarrow H_2O$$

But this equation doesn't balance, because there are 2 oxygen atoms on the left and only 1 on the right. To make things equal, we need to place coefficients of 2 in front of H_2 and H_2O so there are 2 hydrogen atoms and 2 oxygen atoms on each side:

$$2H_2 + O_2 \rightarrow 2H_2O$$

This is known as "balancing by inspection."

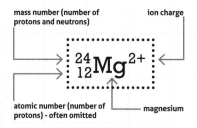

mass number (number of protons and neutrons)

ion charge

$^{24}_{12}Mg^{2+}$

atomic number (number of protons) - often omitted

magnesium

Formula for Success

In the new system, compounds could be shown by putting signs together, and Berzelius introduced the convention of having each sign

Electrolysis

Electrolysis means "breaking apart with electricity," and with the advent of the voltaic pile, this phenomenon became a powerful new tool in the hands of analytical chemists. An electrolytic or electrochemical cell can be used to separate ions, produce redox and displacement reactions, break apart compounds, and isolate pure elements.

Powering Reactions

Electrolysis is a way of passing an electric current to produce chemical reactions at electrodes immersed in an electrolyte. Electrodes are solids, usually strips of metal, which are connected to a source of electricity such as a battery or voltaic cell. Just as the battery has positive and negative terminals, so the electrodes are positive or negative. The positive electrode is called the anode, and the negative electrode is the cathode. An electrolyte is a conductive solution or liquid with ions that can conduct electricity. A typical electrolyte is brine (salt solution). In brine, sodium chloride splits into sodium cations (ions with a positive charge) and chloride anions (ions with a negative charge), and these migrate through the solution when attracted toward the electrodes.

When an electric current is flowing, electrons travel to the cathode, giving it a negative charge, so that it attracts cations. Meanwhile the anode is electropositive and attracts anions. Chemical reactions occur at the interface between the electrolyte and the electrode because of the flow of electrons. At the cathode electrons are donated, causing reduction reactions, and at the anode electrons are lost, causing oxidation reactions. So the electrolytic cell is a device for powering redox reactions (see pp. 90–91).

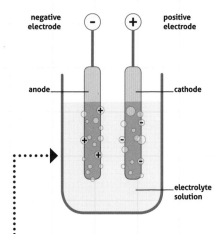

The electrolysis of brine (sodium chloride solution). Reduction and oxidation occur where the electrodes meet the electrolyte. As a result, sodium hydroxide solution is formed, and chlorine gas and hydrogen gas bubble out of the solution.

Electrolysis in Action

Nicholson and Carlisle, the first people to make use of a voltaic pile (see p. 109), managed to decompose water by sticking electrodes of platinum wire into a dish of water. When water is decomposed, hydrogen cations are attracted to the cathode, where they gain electrons and become reduced to hydrogen gas, which bubbles out of solution. At the anode, hydroxyl ions are oxidized to give water and oxygen:

$$4OH^-(aq) \rightarrow O_2(g) + 2H_2O(l) + 4e^-$$

Berzelius used an electrolytic cell to separate salt ions (see pp. 108–109), as occurs when a brine solution is electrolyzed. The chloride anions are attracted to the anode, where they are oxidized to chlorine gas:

$$2Cl^-(aq) \rightarrow Cl_2(g) + 2e^-$$

The Na+ cations migrate to the cathode, but it takes more energy to reduce a sodium ion than a hydrogen ion, so hydrogen gas is produced and the sodium forms sodium hydroxide:

$$2Na^+(aq) + 2H_2O(l) \rightarrow$$
$$H2(g) + 2NaOH(aq)$$

Electroplating

The electrode itself can also take part in electrolysis reactions. For instance, if copper electrodes are immersed in a solution of copper sulfate and a current is passed through it, copper atoms at the

ELECTROLYSIS ON YOUR WRIST

A battery or voltaic pile is basically an electrolysis cell run in reverse, so that electricity is generated rather than used up. A wristwatch battery is an example of a dry cell battery—the zinc outer casing is the anode and a steel cathode is in the center of the battery. The electrolyte is an alkaline paste containing mercury oxide. Most small devices use dry cell batteries; cars use wet cell batteries.

anode will be oxidized to copper cations; and copper cations at the cathode will be reduced to copper atoms that are deposited on the surface of the electrode. Eventually the anode will be eaten away. This process can be used to electroplate the cathode—replace the cathode with a metal object and it will be plated with a thin layer of copper. This is how gold and silver plating is done, and a similar process is used to extract metals from their ores.

An 1882 illustration of early nickel-electroplating apparatus. ∙∙∙∙∙∙∙∙∙∙∙∙

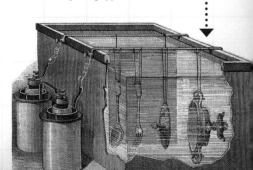

Humphry Davy

The great contemporary of Jöns Jacob Berzelius was the British chemist Humphry Davy, who did more than anyone to raise the profile of his science. Although his contributions to its underlying theory weren't as significant as others, his discoveries and inventions made him the most famous chemist of his era, perhaps the best known of all time.

Happy Gas

Humphry Davy was born in Penzance, Cornwall, England in 1778. He came from a poor and provincial background, yet he was to become highly successful through science. Like Berzelius, he learned chemistry from textbooks, and like many other chemists, he started off as an apothecary's apprentice. In 1798 he took a position at the Pneumatic Institute in Bristol, where the latest pneumatic discoveries were to be applied to medicine.

In 1799 Davy published his first papers, which included an attack on Lavoisier's caloric theory, arguing that heat is motion but light is matter. He first came to public notice through his experiments with nitrous oxide (N_2O), or laughing gas, using himself as a guinea pig and so discovering its psychedelic effects.

Davy's suggestion that the gas could be taken as an anesthetic wasn't followed up for 45 years, but inhaling the gas became fashionable at social gatherings.

Highest Honors

In 1800 Davy showed that the electric current produced by the voltaic pile came from the oxidation of zinc, a finding that led to his election to the Royal Society. In 1801 he became the new star of the Royal Institution, London, where he started a series of popular lectures. Following Lavoisier's prediction that potash and soda were metal oxides, which had so far proved impossible to decompose, he built the most powerful pile yet. Using it to electrolyze their "fused" (molten) states, he managed to isolate pure potassium and sodium. The following year he isolated the alkaline earth metals in the same way.

An 1802 caricature of **experiments with laughing gas** at the Royal Institution, showing Humphry Davy (right) working the bellows.

While researching acids, Davy decomposed hydrochloric acid and found it contained not oxygen but chlorine, which he isolated and named. This disproved Lavoisier's oxygen theory. Davy was knighted in 1812, and after refusing to patent the safety lamp he'd invented (see box), so passing up a fortune, he was made a baron—the highest honor ever given to a scientist. He became President of the Royal Society but his career faded as he became caught up in disputes and social climbing. He spent more time traveling and fishing, dying in Switzerland in 1829.

hen he saw the minute
obules of potassium burst
rough the crust of potash,
d take fire as they entered
e atmosphere… he actually
ounded about the room in
static delight." —**Edmund Davy**

THE LAMP THAT SAVED LIVES

The invention that brought Davy greatest fame was his miner's safety lamp, known as the Davy lamp. In 1815 he was asked to come up with a way of protecting miners from "firedamp"—a build-up of flammable methane, which would explode if it came in contact with the heat from a flame. Davy discovered that if a metal gauze was put around the flame, it quickly absorbed the heat while still allowing light to pass through the holes. The flame was no longer hot enough to ignite the methane. This meant that a cheap and robust safety lamp could be made by enclosing the wick in a mesh cylinder or chimney.

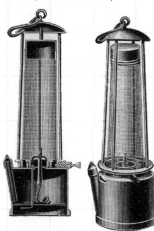

The Davy lamp could also be used as a gas detector, the flame burning higher or lower or changing color depending on the type of gases present.

5

The Periodic Table

The development of inorganic chemistry and the quest for the elements reached a climax with the discovery of the periodic table, a simple structure that brought together the discoveries of the chemical revolution into a logical and unified whole. This chapter explains the principles that underlie the periodic table, tells the story of its discovery and confirmation, and introduces the most important aspects of subsequent developments in chemistry, namely nuclear and organic chemistry.

The Periodic Table

The periodic table in its current form runs to 118 elements, though at the upper range of atomic numbers the elements are highly unstable and may have existed for only fractions of a second in a particle accelerator collision chamber. To prevent the table from becoming unmanageably wide, the "f block" of elements (lanthanides and actinides) is usually taken out of the table and shown as a separate block.

The periodic table opposite shows the 118 known elements, arranged in order of increasing atomic number. Color coding indicates categories of elements that have similar properties. Note that hydrogen (H) is hard to categorize and in some versions of the table it is shown on its own as a block of one.

The table below lists the names and atomic weights of the 109 elements whose names are widely accepted and have been approved by the International Union of Pure and Applied Chemistry (IUPAC).

Widely accepted element names

Ac Actinium 227	**Au** Gold 196.9665	**Br** Bromine 79.904	**Cm** Curium 247	**Ds** Darmstadtium 278	**Fm** Fermium 257	**Hf** Hafnium 178.49	**K** Potassium 39.0983
Ag Silver 107.8682	**B** Boron 10.811	**C** Carbon 12.0107	**Cn** Copernicium 285	**Dy** Dysprosium 162.5	**Fr** Francium 223	**Hg** Mercury 200.59	**Kr** Krypton 83.8
Al Aluminum 26.9815	**Ba** Barium 137.327	**Ca** Calcium 40.078	**Co** Cobalt 58.9332	**Er** Erbium 167.259	**Ga** Gallium 69.723	**Ho** Holmium 164.9303	**La** Lanthanum 138.9055
Am Americium 243	**Be** Beryllium 9.0122	**Cd** Cadmium 112.411	**Cr** Chromium 51.9961	**Es** Einsteinium 252	**Gd** Gadolinium 157.25	**Hs** Hassium 277	**Li** Lithium 6.941
Ar Argon 39.948	**Bh** Bohrium 264	**Ce** Cerium 140.116	**Cs** Cesium 132.9055	**Eu** Europium 151.964	**Ge** Germanium 72.64	**I** Iodine 126.9045	**Lr** Lawrencium 262
As Arsenic 74.9216	**Bi** Bismuth 208.9804	**Cf** Californium 251	**Cu** Copper 63.546	**F** Fluorine 18.9984	**H** Hydrogen 1.0079	**In** Indium 114.818	**Lu** Lutetium 174.967
At Astatine 210	**Bk** Berkelium 247	**Cl** Chlorine 35.453	**Db** Dubnium 262	**Fe** Iron 55.845	**He** Helium 4.0026	**Ir** Iridium 192.217	**Md** Mendelevium 258

Legend

- alkali metals
- transition metals
- other non-metals
- alkaline earth metals
- other metals
- halogens
- lanthanides
- metalloids
- noble gases
- actinides

The Periodic Table

1																	18
1 H	2											13	14	15	16	17	2 He
3 Li	4 Be											5 B	6 C	7 N	8 O	9 F	10 Ne
11 Na	12 Mg	3	4	5	6	7	8	9	10	11	12	13 Al	14 Si	15 P	16 S	17 Cl	18 Ar
19 K	20 Ca	21 Sc	22 Ti	23 V	24 Cr	25 Mn	26 Fe	27 Co	28 Ni	29 Cu	30 Zn	31 Ga	32 Ge	33 As	34 Se	35 Br	36 Kr
37 Rb	38 Sr	39 Y	40 Zr	41 Nb	42 Mo	43 Tc	44 Ru	45 Rh	46 Pd	47 Ag	48 Cd	49 In	50 Sn	51 Sb	52 Te	53 I	54 Xe
55 Cs	56 Ba	57–71 ☆ lanthanides	72 Hf	73 Ta	74 W	75 Re	76 Os	77 Ir	78 Pt	79 Au	80 Hg	81 Tl	82 Pb	83 Bi	84 Po	85 At	86 Rn
87 Fr	88 Ra	89–103 ☆☆ actinides	104 Rf	105 Db	106 Sg	107 Bh	108 Hs	109 Mt	110 Ds	111 Rg	112 Cn	113 Uut	114 Uuq	115 Uup	116 Uuh	117 Uus	118 Uuo

☆ lanthanides

57 La	58 Ce	59 Pr	60 Nd	61 Pm	62 Sm	63 Eu	64 Gd	65 Tb	66 Dy	67 Ho	68 Er	69 Tm	70 Yb	71 Lu

☆☆ actinides

89 Ac	90 Tn	91 Pa	92 U	93 Np	94 Pu	95 Am	96 Cm	97 Bk	98 Cf	99 Es	100 Fm	101 Md	102 No	103 Lr

Mg Magnesium 24.305
Nd Neodymium 144.24
P Phosphorus 30.9738
Pt Platinum 195.078
Rh Rhodium 102.9055
Sg Seaborgium 266
Tc Technetium 98
V Vanadium 50.9415

Mn Manganese 54.938
Ne Neon 20.1797
Pa Protactinium 231.0359
Pu Plutonium 244
Rn Radon 222
Si Silicon 28.0855
Te Tellurium 127.6
W Tungsten 183.84

Mo Molybdenum 95.94
Ni Nickel 58.6934
Pb Lead 207.2
Ra Radium 226
Ru Ruthenium 101.07
Sm Samarium 150.36
Th Thorium 232.0381
Xe Xenon 131.293

Mt Meitnerium 268
No Nobelium 259
Pd Palladium 106.42
Rb Rubidium 85.4678
S Sulfur 32.065
Sn Tin 118.71
Ti Titanium 47.867
Y Yttrium 88.9059

N Nitrogen 14.0067
Np Neptunium 237
Pm Promethium 145
Re Rhenium 186.207
Sb Antimony 121.76
Sr Strontium 87.62
Tl Thallium 204.3833
Yb Ytterbium 173.04

Na Sodium 22.9897
O Oxygen 15.9994
Po Polonium 209
Rf Rutherfordium 261
Sc Scandium 44.9559
Ta Tantalum 180.9479
Tm Thulium 168.9342
Zn Zinc 65.39

Nb Niobium 92.9064
Os Osmium 190.23
Pr Praseodymium 140.9077
Rg Roentgenium 283
Se Selenium 78.96
Tb Terbium 158.9253
U Uranium 238.0289
Zr Zirconium 91.224

Periodic Table Pioneers

The rush of elemental discoveries of the early 19th century, and the explanation of atomic weight and the law of definite proportions, all seemed to be building toward something: a grand fusion that would unite the microcosmic world as Newton's laws of gravity had united the macrocosmic world. But who would achieve this epoch-making breakthrough?

The fusion that chemistry was building toward was the periodic table, brainchild of the great Russian chemist Dmitry Mendeleyev (see pp. 122–123). But before its discovery there were three important prototypes—theories that glimpsed a part of the whole that Mendeleyev would reveal in 1869, but which failed because chemistry was incomplete at this time.

Law of Triads

The first of these periodic table pioneers was German chemist Johann Wolfgang Döbereiner (1780–1849). He noted that the recently discovered element bromine not only had properties that seemed intermediate between chlorine and iodine, but it also had an atomic weight that seemed to be midway between the two. Studying the rest of the elements, he spotted two more of these groups of three, which he called "triads": calcium—strontium—barium, and sulfur—selenium—tellurium. In 1829 he announced his "law of triads," but as it seemed to apply to only 9 of the 54 elements known at the time, it attracted little attention.

Johann Wolfgang Döbereiner (1780–1849).

Alkali formers

symbol	A (atomic mass)
Li	7
Na	23
K	39

Salt formers

symbol	A (atomic mass)
Cl	35.5
Br	80
I	127

The Telluric Screw

In 1860 Avogadro's theories were finally accepted, leading to a revised and more accurate table of atomic weights. French geologist Alexandre-Émile Béguyer de Chancourtois (1820–1886) was the first person to list the elements in order of these revised weights. He plotted the weights in a spiral on the outside of a cylinder and noticed that elements with similar properties lined up in vertical columns. He called this pattern the *vis tellurique*, or "telluric screw," as tellurium was at the center of his system. Unfortunately, when Chancourtois published his paper in 1862, the journal left out his explanatory diagram, making it almost impossible for readers to visualize his spiral arrangement. Unsurprisingly, few people took any notice.

Law of Octaves

Just two years later the English chemist John Newlands (1837–1898) listed the elements in ascending order of atomic weight in rows of seven. He found that this pattern gave rows of elements with similar properties: each element was similar to the element eight places ahead of it. Being a fan of musical theory, he likened this to the eighth note in an octave and called his sequence the "law of octaves." Newlands reported his theory in a paper in 1865, but there were many holes in his scheme—especially at higher atomic weights, where the pattern broke down—and he was met with ridicule. After Mendeleyev's system was published, Newlands claimed priority, though his scheme lacked the innovations that made the Russian's so brilliant. Eventually the Royal Society awarded Newlands the Davy Medal in 1887.

De Chancourtois's explanatory graph, drawn up to illustrate his telluric screw concept. Published without this, his 1862 paper was ignored.

Part of Newlands's table

H	Li	Be	B	C	N	O
F	Na	Mg	Al	Si	P	S
Cl	K	Ca	Cr	Ti	Mn	Fe

Part of Newlands's table, in which he listed the elements in ascending order of atomic weight in rows of seven.

Dmitry Mendeleyev

Hailed as the greatest chemical mind since Lavoisier, Dmitry Mendeleyev did important work for industrial and agricultural chemistry, helped regulate Russia's weights and measures, and authored a landmark textbook. But his most enduring achievement was his periodic law or, as it is better known, the periodic table, which ranks alongside the achievements of Charles Darwin and Isaac Newton.

Visionary Dream

Dmitry Mendeleyev (1834–1907) was born in Siberia, the youngest in a large family. A brilliant student, he overcame illness to win a scholarship to study in Germany with Robert Bunsen (see pp. 130–131), but later returned to Russia to take up a position at St. Petersburg University.

While working on a new textbook in 1869, Mendeleyev was prompted to consider whether the elements could be arranged according to some system or law. He was familiar with the work of de Chancourtois and began to play with the order of the elements for himself. He noted that the halogens and the oxygen and nitrogen groups could be arranged in a table of ascending atomic weights. Seeking a larger pattern that included all the other elements, he wrote the name and atomic weight of each one on a card and arranged them in vertical lines. After working on the problem for three days, he fell asleep and had a famous dream: "I saw in a dream a table where all the elements fell into place…"

A Daring New System

Mendeleyev's paper "A Suggested System of the Elements" showed a table in which the elements were ordered in columns of descending atomic weight, arranged such that each row contained elements with similar properties. What was revolutionary and daring about his scheme was that it wasn't confined by the constraints that had crippled previous efforts. Where necessary he put some elements out of order, writing question marks next to their atomic weights, and left gaps where there was no element that fitted the pattern.

The true test of a scientific theory is that it makes testable predictions, and Mendeleyev's periodic law did just this. Not only was he able to predict which atomic weights had probably been incorrectly established; he was able to predict the existence of as yet unknown elements, including their likely atomic weights and even their properties. These unknown elements included eka-aluminum with a predicted atomic weight of 68 and eka-silicon with a predicted atomic weight of 70 (*eka* means "one" in Sanskrit).

There was a connection between each of his horizontal rows, or groups, and the valence (combining power) of the elements it contained, which confirmed his belief that the table was accurate. Reading vertically along the table, the valencies went from 1 on the lithium row up to 4 on the carbon row and back down to 1, giving a pattern of 1, 2, 3, 4, 3, 2, 1—a periodic (recurring) rise and fall. Here was the periodic law he'd been looking for. Although there were inconsistencies, he was confident enough to overlook them.

The original **Russian version of Mendeleyev's periodic table**, with periods running vertically instead of horizontally. Notice the question marks next to elements that he predicted but which hadn't yet been discovered.

This **monument to the periodic table** can be found at the Slovak University of Technology in Bratislava, Slovakia. Mendeleyev's distinctive portrait occupies center stage.

"I have never doubted the universality of this law, because it could not possibly be the result of chance." —**Dmitry Mendeleyev**

Periodic Law

The periodic law that Mendeleyev had uncovered, though refined in subsequent years, is the key to inorganic chemistry. With this law chemists can make sense of both the big picture and the fine details of their field, arranging the elements into groups with similar physical and chemical properties, and predicting how they will interact and even which ones have yet to be discovered.

Ordering the Elements

Like the periodic pioneers before him, Mendeleyev ordered the elements in his system according to their atomic weight. At the time, the concept of subatomic particles was only speculation and there was no way of knowing whether protons existed, let alone count them. This caused problems for the new periodic system, because atomic weight doesn't affect the chemistry of an element. It's the number of electrons in the outer (valence) shell that gives an element its chemical properties (see pp. 28–29), and this in turn is decided by the number of protons—its atomic number. For this reason, the modern periodic table is ordered by atomic number, as its main function is to show elements with similar chemical behavior.

The table sorts the elements into seven rows, or "periods," with atomic numbers increasing along the row from left to right. This arranges the elements into columns called "groups," and in each group the elements have similar physical and chemical properties.

Shown below are the first four periods consisting of 2, 8, 8, and 18 elements respectively. What's the link between this order and the repeating patterns in chemical and physical properties? These numbers tell us the size of valence shell of

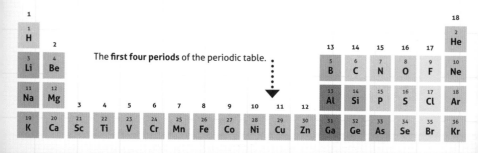

The **first four periods** of the periodic table.

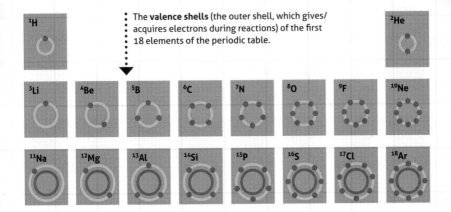

The **valence shells** (the outer shell, which gives/acquires electrons during reactions) of the first 18 elements of the periodic table.

each period. The first period contains hydrogen and helium. These elements have only 1 electron shell as their valence shell, which can only contain 2 electrons. The next electron shell available can have up to 8 electrons, as can the third, while the fourth can hold up to 18. Shown above are the valence shell configurations of the first 18 elements in the table. Actually the picture is slightly more complicated than described here, as the shells are subdivided into orbitals (see p. 129).

Predictive Power

As Mendeleyev proved, periodic law is a powerful tool that allows chemists to predict the potential existence of elements not yet discovered. Several years went by without anyone discovering eka-aluminum and eka-silicon, two elements Mendeleyev had predicted

to exist. But French chemist Paul Lecoq de Boisbaudran (1838–1912) was determined to find one of them. He knew eka-aluminum would have an atomic weight of about 68, so he looked for it in an ore of zinc, which has an atomic weight of around 65. He eventually identified a new element using spectroscopy (see pp. 130–131). It had an atomic weight of 69.72 and he named it gallium.

Further discoveries offered yet more confirmation that Mendeleyev was right. In 1879 the element scandium was discovered, matching his prediction of eka-boron, and in 1886 his eka-silicon was found, and named germanium.

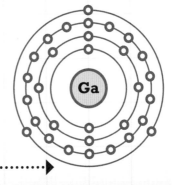

The **electron configuration of gallium (Ga)**, showing how it has three electrons in its outer orbital. This gives it a valence of 3.

Understanding the Table

The periodic table arranges elements into categories and groups, making it much simpler to classify and describe them. Understanding these group relationships is essential for any chemistry student. Grasping the underlying rules that determine group characteristics makes chemistry less complex, helping to make sense of the confusion of names and terms.

Metals, Non-metals, and Metalloids

There are various ways of dividing up the elements in the table. One is to split them into three broad categories—metals, non-metals, and metalloids. The metals are all the elements to the left of a stepped line drawn down the table, starting at element 5, boron (B), and extending down to element 84, polonium (Po), with the exception of germanium (Ge) and antimony (Sb). The non-metals (together with hydrogen) are to the right of the line, while the elements bordering the line are the metalloids.

Metals have physical properties that are familiar from everyday life. They are almost all solids, most are hard, dense and shiny, and give a "ping" if struck. They're ductile and malleable (they can be drawn into thin wires or hammered into flat sheets). Chemists classify a metal by conductivity— metals are good conductors of heat and electricity. In chemical reactions they generally lose, or donate, electrons.

Non-metals include several gases and liquids; when solid they tend to be brittle. They are poor conductors and in chemical reactions they tend to gain electrons.

Metalloids, also known as semimetals, combine characteristics of the other two groups, including conductivity. As a result they are semiconductors, which are widely used in electronics.

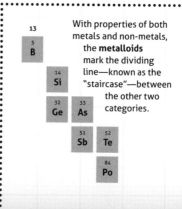

With properties of both metals and non-metals, the **metalloids** mark the dividing line—known as the "staircase"—between the other two categories.

Most of the elements in the periodic table form oxides. How these behave in water is generally determined by whether the element is metallic or non-metallic. Metals form basic oxides—they react with water to form alkaline solutions. Non-metals form acidic oxides—they react with water to give acid solutions. Some elements, such as aluminum, form amphoteric oxides, meaning they can go either way.

Group Connections

Another way of dividing up the periodic table is into its vertical columns, which form groups. These are numbered from 1 to 18—or, more traditionally, with Roman numerals and letters. It's worth looking in greater detail at some notable groups:

Alkali Metals (group 1)

The alkali metals are highly reactive and soft enough to cut with a knife. They have one outermost or valence electron, which they lose to form ions with a single positive charge and so their oxidation state is +1 (see pp. 90–91). Hydrogen is placed on the table as if it were a member of group 1, but this is only because of its atomic number. It's actually in a group of its own.

1			
3 **Li**			
11 **Na**	**Li** Lithium 6.941	**K** Potassium 39.0983	**Cs** Cesium 132.9055
19 **K**	**Na** Sodium 22.9897	**Rb** Rubidium 85.4678	**Fr** Francium 223
37 **Rb**			
55 **Cs**			
87 **Fr**	The six elements that make up the **alkali metals**.		

Alkaline Earth Metals (group 2)

The alkaline earth metals are also generally very reactive. Like the Group 1 metals, they are found in nature as ionic salts, but because they each have two valence electrons, they have an oxidation state of +2.

2			
4 **Be**			
12 **Mg**	**Be** Beryllium 9.0122	**Ca** Calcium 40.078	**Ba** Barium 137.327
20 **Ca**	**Mg** Magnesium 24.305	**Sr** Strontium 87.62	**Ra** Radium 226
38 **Sr**			
56 **Ba**			
88 **Ra**	The six elements that form the **alkaline earth metals**.		

Halogens (group 17)

The halogens get their name from their tendency to react with metals to form salts (*halx* in Greek). They each have seven valence electrons so they tend to be strong oxidants, gaining one electron to form ions with a single negative charge.

17		
9 **F**		
17 **Cl**	**F** Fluorine 18.998	**Cl** Chlorine 35.453	**Br** Bromine 79.904
35 **Br**	**I** Iodine 126.905	**At** Astatine 210	
53 **I**		
85 **At**	The five elements that make up the **halogens**.	

All three of these groups display trends typical of periodic groups. Their characteristic group properties tend to get weaker as you go down the column, though the first member of each group is often slightly atypical—for example, the chemistry of lithium is different from the other alkali metals.

Noble Gases (group 18)

The noble gases weren't known at the time Mendeleyev constructed his table, and when they were discovered, he was at first troubled that a new class of elements would destroy his theory. In fact they proved to be the final piece of the puzzle, slotting in neatly at the end of the table. They have complete valence shells of eight electrons, making them extremely unreactive.

18		
2 **He**		
10 **Ne**	**He** Helium 4.003	**Ne** Neon 20.179	**Ar** Argon 39.948
18 **Ar**	**Kr** Krypton 83.798	**Xe** Xenon 131.293	**Rn** Radon 222
36 **Kr**		
54 **Xe**		
86 **Rn**	The six naturally occurring **noble gases** all have very low reactivity.	

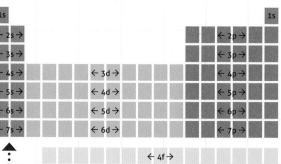

As shown in this chart, the periodic table can be broken into **blocks**. The blocks correspond to the highest energy electrons.

Letter Blocks

....

A third way of dividing up the table is by electron orbitals. As explained on pages 124–125, each period represents electron shells successively more distant from the nucleus. These shells are subdivided into s, p, d, and f orbitals, which contain up to 2, 6, 10, and 14 electrons respectively. As atomic number increases across each period, so elements begin to fill each of these orbitals in turn, and the table can be split into blocks corresponding to these orbitals:

The **s block** contains groups 1 and 2.

The **p block** includes groups 13 to 18—moving from left to right along these periods, the p orbital is progressively filled.

The **d block** includes groups 3 to 12, known as the transition metals—as you move across the periods, electrons progressively fill the d orbital. Because the d orbital can hold up to 10 electrons, the d block is 10 elements across.

The **f block** is usually shown as a separate block below the table to save space, though the f orbital kicks in when you get to periods 6 and 7. It includes elements known as the lanthanides or rare earths (on period 6), and the actinides (on period 7), which are all radioactive elements. Because the f orbital can hold up to 14 electrons, the f block is 14 elements across.

SUPER-HEAVY ELEMENTS

At very high atomic numbers, the nucleus of the atom becomes very large and unstable, and the elements become radioactive (see pp. 132-133), decaying (breaking down) to give elements with lower atomic numbers. This means that the highest naturally occurring elements are element 92, uranium (U) and 94, plutonium (Pu). If elements heavier than this happen to form, perhaps in a nuclear reaction, they will almost instantly decay. However, atom-smashing technology has allowed scientists to artificially create many of the super-heavy elements predicted by the periodic table. The most recently created element, at the time of writing, is element 117, ununseptium (Uus). A US-Russian team achieved this in 2010 by smashing together atoms of element 20, calcium (Ca) and 97, berkelium (Bk). The heaviest element ever created is 118, ununoctium (Uuo). The names for these super-heavy atoms are temporary, created by the International Union of Pure and Applied Chemistry (IUPAC) until permanent names and symbols can be agreed.

Measuring Light: Spectroscopy

As periodic law reveals, chemistry and physics are intimately connected. The voltaic pile, a technology born of chemistry, had signaled the 19th century and the start of the modern age by opening new worlds in chemistry and physics. Now a new light-measuring technology born of the marriage between chemistry and physics, would extend the scope of both literally to new worlds—the stars!

Studying Sunlight

The German optician Joseph von Fraunhofer was the first to notice that the spectra (range of wavelengths) of light from a flame had distinct lines of brightness when viewed through optical glass. Directing his glasses at the Sun, he discovered that its continuous light spectrum was broken up by dark lines, which remained something of a mystery.

Solar spectrum with Fraunhofer's lines.

It was at this time that Robert Bunsen, professor of chemistry at The University of Heidelberg, Germany, noticed that burning some elements produced flames with characteristic colors. Heidelberg's professor of physics Gustav Kirchhoff helped Bunsen analyze the problem. In 1859 they used a prism to separate the different wavelengths by refraction (making them change direction) and obtained spectra from different elements. They also proved that each one emitted a unique spectrum that could be used to identify it—a spectral fingerprint.

Known as "spectrochemical analysis," in 1860–61 this technique allowed Bunsen to prove the existence of two previously unknown elements found in mineral water in trace amounts—including one that produced deep red spectral lines and was therefore named rubidium (from the Latin for "deep red"). Meanwhile, Kirchhoff ingeniously applied the technique in reverse to analyze sunlight, and he and Bunsen were able to show that von Fraunhofer's dark lines matched up with the bright yellow lines produced by sodium. They concluded that sodium must be present in the atmosphere of the Sun—and rather than emitting yellow light, it was absorbing it and blocking those portions of the spectrum from reaching Earth. Science had now made it possible to analyze the elements of the stars.

Quantum Leaps
······················

The analysis of spectra that a substance emits is known as "emission spectroscopy." This depends on the rules governing the energy levels, or states, of electrons orbiting an atom. When an electron absorbs a packet of light energy known as a quantum, it can jump from its resting or ground state to an excited state. When it falls back, it releases the quantum, emitting it as light. The number of quanta needed depends on the orbitals in question. In turn this determines the energy and therefore wavelength of light emitted when the electron falls back from its excited state to its ground state. Each element has a unique configuration of electrons, and therefore a unique absorption and emission spectrum. This means that the spectrum can be used to read the electron configuration and the atomic number of the element.

LIGHT INSTRUMENTS

The basic instrument of spectroscopy is the spectroscope. It has a slit to allow only a beam of light through, a collimator (a device for narrowing a beam of light so that its rays are parallel), a prism or grating to separate wavelengths by refraction or diffraction (they spread out), and a telescope or microscope objective lens for the user to look through. Adding a camera or other recording device to the instrument makes it into a spectrograph, while adding a calibrated scale for measuring the spectra makes it a spectrometer.

A simplified representation of a **spectrograph**, with a beam from a collimator (entering from the right) split into a spectrum by a prism and focused onto a photographic plate.

Isotopes and Radioactivity

Normal chemistry deals almost exclusively with the action and behavior of electrons. The nuclei of atoms hardly get a look in, as for the most part they're involved in the phenomena of covalent and ionic bonding, electrochemistry, and the like. When the nucleus does get involved it's termed "nuclear chemistry," which deals with isotopes, radioactivity, and nuclear reactions.

Isotopes

Isotopes are atoms with the same atomic number but with more or fewer neutrons and therefore different atomic weight. For instance, carbon-14 and carbon-12 both have 6 protons and so both have the same atomic number. However, carbon-14 has 8 neutrons in its nucleus, and carbon-12 has only 6, giving them atomic masses of 14 and 12 respectively. Because both isotopes have the same number of protons, they also have the same number of electrons and so the same chemistry. Another example is uranium-238 and uranium-235. Both isotopes have the atomic number 92, but uranium-238 has 146 neutrons and uranium-235 only 143. The number of neutrons in an isotope equals atomic mass minus atomic number. In scientific notation, an isotope is shown by giving the atomic mass as a superscript in front of the element symbol:

^{14}C and ^{12}C; ^{238}U and ^{235}U

Isotopes aren't found in equal abundance on Earth. The vast majority of carbon atoms, for instance, are made up of ^{12}C. This means an element's "average" atomic mass—or atomic weight— is based on the most abundant isotope. So the atomic weight of carbon is very close to 12 (atomic mass equals 12.0115). Of the 83 elements that occur naturally in significant quantities on Earth, 20 are found as a single isotope (mononuclidic), and the others as mixtures of up to 10 isotopes.

Radioactive Decay

Radioactivity is the decay of an unstable nucleus, involving the loss and/or transformation of subatomic particles with release of energy. The stability of the nucleus depends on the proton:neutron (P:N) ratio. If an isotope has too few or too many neutrons, it will be unstable—the stable ratio depends on the atomic number. Nuclei also become unstable when they get above a certain size—all the elements with an atomic number of 84 or more are unstable and therefore radioactive. Radioactive decay takes places when the nucleus "wants" to achieve a more stable P:N ratio. This can involve three types of radiation—alpha particles, beta articles, and gamma rays:

Alpha particles have 2 protons and 2 neutrons. When an atom emits an alpha particle, its atomic mass falls by 4 atomic mass units (amu) and its atomic number by 2. Alpha particle emission is typical of heavy elements such as uranium. Decay is a form of nuclear reaction and can be described in an equation similar to that used in chemical reactions. For example:

$$\text{[atomic mass]} \quad {}^{238}_{92}U \rightarrow {}^{234}_{90}Th + {}^{4}_{2}He$$
$$\text{[atomic number]}$$

A **beta particle** is an electron that is emitted from the nucleus when a neutron decays into a proton plus an electron. The electron shoots out of the nucleus, leaving the proton behind. This means the atomic mass doesn't change but the atomic number increases by 1. For example, the hydrogen isotope known as tritium has 2 neutrons and 1 proton. This is an unstable P:N ratio, so one of the neutrons decays to emit a beta particle, becoming a proton in the process. This changes the atomic number to 2, converting the hydrogen atom into an isotope of helium:

$${}^{3}_{1}H \rightarrow {}^{3}_{2}He + {}^{0}_{-1}e \text{ [beta particle]}$$

A DREAM ACHIEVED

Radioactive decay is a form of transmutation, because it accomplishes what the alchemists always dreamed of—changing one element into another. But contrary to what they thought, naturally occurring transmutation is more likely to change a valuable element into a base metal than vice versa—uranium, for instance, decays in a series of steps until it becomes lead. Artificial transmutation is possible with atom smashers, which can add protons and neutrons to elements.

Even though the beta particle is just an electron, it's written with a specific notation to balance the equation. Just as with a chemical equation, the numbers on both sides must match up—in a nuclear reaction, the atomic mass and numbers must match up. In this case:

$(3 = 3 + 0)$ and $(1 = 2 + -1)$.

Gamma radiation is a form of electromagnetic energy. Sometimes a nucleus that has undergone alpha or beta decay is left in an excited state and drops down to a lower energy state by emitting a very high-energy photon, known as a gamma ray. Gamma rays are close to X-rays on the electromagnetic spectrum.

Marie and Pierre Curie

The key figures in the explanation of radioactivity were Marie and Pierre Curie. Marie has become particularly celebrated; the first person to win two Nobel prizes, she overcame hardship and prejudice to blaze a trail for women scientists. The Curies' pioneering research helped reveal the chemistry of radioactivity.

Combining Talents

Marie Curie (1867–1934) was born in Poland, the youngest daughter of poor school teachers. She struggled to get an education and worked as a governess to help pay for her sister's medical education in Paris. In 1891 she was able to join her sister and studied at the Sorbonne, where she met French chemist Pierre Curie (1859–1906). Pierre had discovered the "piezoelectric effect," where crystals generate a charge if put under pressure, and concepts influencing the magnetic properties of substances.

He and Marie married in 1895, and she used instruments designed by him to pursue her dissertation—an examination of the uranium-bearing ore, pitchblende. The French physicist Henri Becquerel (1852–1908) had found that uranium gave off radiation, and Marie hoped to find similar rays being given off by pitchblende. It turned out to be even more strongly radioactive, suggesting that it might contain other, undiscovered radioactive elements. Her husband abandoned his own research and joined her in the exhausting task of isolating new elements from huge quantities of pitchblende.

Radioactive Discoveries

In 1898 the Curies discovered element 84, polonium (Po), named for Marie's homeland, and 88, radium (Ra). They coined the term "radioactivity" and proved that beta

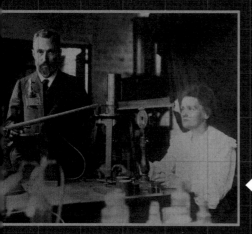

French chemist **Pierre Curie** (1859–1906) and Polish-born **Marie Curie** (1867–1934).

radiation involved particles with a negative charge, laying the groundwork for understanding atomic structure. In 1903 the Curies shared the Nobel Prize in Physics with Becquerel, but just three years later Pierre died and Marie took over his teaching post, becoming the first woman to teach at the Sorbonne. During World War I she helped direct medical use of radioactive elements, and in 1916 was awarded the Nobel Prize in Chemistry for her work on radium. However, her health was broken by long exposure to dangerous substances and she died of leukemia.

The discovery of polonium and radium led others to isolate further radioactive elements, revealing a whole decay series from uranium to lead. This explained why pitchblende contained so many

HALF-LIFE AND CARBON DATING

Carbon-14 (14C) is a radioactive isotope with a half-life of 5,730 years, and this can be used to date organic remains up to about 60,000 years old. Carbon-14 is found in carbon dioxide molecules in the air. This enters the food chain, which means there's an amount of 14C in all living organisms. As soon as an organism dies, the amount of 14C it absorbed gradually decays—half of it every 5,730 years. By comparing how much 14C is left in the dead organism with the amount in a living organism, it's possible to estimate when the organism died.

radioactive elements: one was formed by radioactive decay of the next. It was impossible to predict when a single radioactive atom would decay, but with a large enough sample, scientists could give a possible answer by saying how long it would take for half the atoms in a sample to decay. This figure is known as the half-life. For instance, the half-life of radon-222 is 3.8 days—so after 3.8 days, half a sample of Rn-222 will have decayed, and after 7.6 days there will only be a quarter as many atoms.

Pitchblende, now known as uraninite. Marie and Pierre Curie found that this mineral not only contained uranium, but also two new elements which Marie called polonium and radium. It also contains many other radioactive elements, such as radium and lead, which can be traced back to the decay of uranium.

Organic Chemistry: A Very Brief Introduction

Organic chemistry is the chemistry of carbon, an element with unique properties that has created a whole sphere of science. Because of these properties, the chemistry of carbon is also the chemistry of life, and everything that derives from it, from petrol to plastics. Here's a short explanation of the main concepts and terms.

Discovering the Carbon Skeleton

A carbon atom has six protons and therefore six electrons—two in its inner shell and four in its outer (valence) shell. The four electrons in the valence shell are the key to carbon's properties, allowing a carbon atom to form four covalent bonds (sharing an electron pair) with other atoms, including other atoms of carbon. These bonds can be single, double, or triple. Self-bonding means that carbon can form long chains, that can act as skeletons to which other elements attach. The number of possible combinations of carbon atoms and their attachments is basically limitless.

The enormous diversity and complexity of organic chemistry posed a huge challenge for early chemists, once they'd begun to distinguish between organic and inorganic substances in the late 18th century. Lavoisier showed that the components of organic compounds were actually very limited—all included carbon and hydrogen, often with oxygen and occasionally nitrogen. But as research into organic chemistry

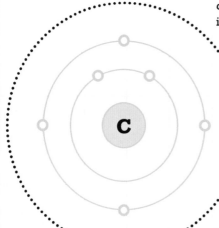

The electronic configuration of a carbon atom. A carbon atom has six protons and six electrons—two in its inner shell and four in its outer (valence) shell. The four in the outer shell allow it to form four covalent bonds.

progressed, it became harder to impose any systematic order. At least one pioneer in the field, the German chemist Justus von Liebig (1803–1873), became so exasperated that he gave up on trying to systematize and turned to applied organic chemistry instead. Not until 1858 was August Kekulé (1829–1896) able pull together all the research and compile a comprehensive theory of chemical structure, emphasizing the importance of carbon backbones or skeletons.

Hydrocarbons

The simplest organic compounds are created when only hydrogen atoms are attached to this carbon skeleton. These are called "hydrocarbons," and even these are extremely diverse. The naming system for hydrocarbons is based on the type of bonds between atoms in the carbon chain:

Alkanes: molecules with only single bonds. In an alkane every carbon atom makes four bonds to four different atoms, and these molecules are said to be "saturated."

Alkenes: molecules with one or more double bonds.

Alkynes: molecules with one or more triple bonds.

Cyclic hydrocarbons or **cyclohexenes:** molecules in which carbon atoms link up in a ring—because the rings are made up of six carbon atoms. An important class of cyclohexenes is the **aromatics**, where the cyclohexene ring has alternating single and double bonds.

Alkanes and Alkenes

Hydrocarbon type	Molecule	Formula	Chemical structure	Molecular model
Alkane	Methane	CH_4	H / H–C–H / H	
Alkane	Ethane	C_2H_6	H H / H–C–C–H / H H	
Alkene	Ethene	C_2H_4	H H / C=C / H H	
Alkene	Propene	C_3H_6	H H H / H–C–C=C / H H	

Many organic compounds are too complex to be described by their molecular (chemical) formula, because of the potential for double and triple bonds and for branching of carbon chains. So they also have a structural formula, showing how the atoms are arranged in the molecular structure. For instance, the hydrocarbon butane has the molecular formula C_4H_{10}, but it can take on one of two forms with different structural formulae. In normal butane, the structural formula shows the carbon atoms in a straight chain:

$$CH_3-CH_2-CH_2-CH_3$$

This is a condensed structural formula—different from an expanded one, which shows each hydrogen atom separately and all the bonds between each atom:

Carbon atoms on the end of the chain have three spare bonds and therefore bond to three hydrogen atoms. Those in the middle of the chain use up two bonds linking to the carbon atoms on either side, and therefore bond to only two hydrogen atoms.

Another arrangement is possible, with one carbon atom branching off the chain:

$$CH_3-CH-CH_3$$
$$\vert$$
$$CH_3$$

A compound with the same molecular formula but different structural formula is known as an isomer, so this is known as isobutane.

Functional Groups

When an element other than hydrogen bonds with an organic molecule, it's known as a functional group. Important functional groups include the alcohols, where an –OH group is bound to the carbon backbone; and the amines, where a nitrogen-containing group, $-NH_2$, is involved. The simplest form of alcohol is methanol (also known as methyl or wood alcohol): CH_3OH. Ethanol, the alcohol found in wine, beer, and spirits, is CH_3CH_2OH.

Wine, beer, and spirits contain **ethanol**, a form of alcohol.

KEKULÉ'S SNAKY DREAM

August Kekulé famously claimed that the hexagonal ring-shaped structure of the simplest cyclohexene, benzene (C_6H_6), came to him in a dream. While trying to understand benzene's structure he dozed off, picturing atoms twisting in a snakelike motion. Seeing that "one of the snakes had seized hold of its own tail," he suddenly saw the shape of the structure.

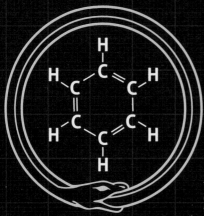

The atomic **structure of benzene evokes the image of a snake eating its own tail.**

THE ELEMENT OF LIFE

Of all the elements in the universe, carbon is the fourth most abundant and the most important for life. All living things contain carbon in some form. It is the building block of life because it can form so many different kinds of bonds and form essential compounds. On Earth, the carbon cycle allows such processes as photosynthesis, respiration, decomposition, and carbonification (the conversion of vegetable matter to coal).

Carbon plays a vital role in photosynthesis.

Index
·········

GLOSSARY

················

*Terms are explained where they
are introduced within the text.
However, a few are noted here
for the sake of clarity.*

Acid a compound that raises
concentration of H+ ions in water.

Activation energy an energy boost
that gets a reaction started.

Alkali a base that dissolves in water
to produce hydroxide ions.

Atomic mass the mass of an atom
in atomic mass units (amu).

Atomic weight the average of an
element's isotope masses, taking
into account the isotopes' natural
abundance on Earth.

Base a compound that reacts with
an acid to produce a salt.

Combustion reaction where a
compound combines with oxygen;
otherwise known as burning redox
(short for reduction-oxidation).

Covalent bond a bond formed when
two atoms share a pair of electrons;
the electrons effectively take up a
new orbit encompassing both atoms.

Endothermic reaction a reaction
that absorbs heat energy from the
surroundings.

Exothermic reaction a reaction that
generates heat.

Inorganic chemistry the chemistry
of the elements and the compounds
of elements other than carbon.

Ion an atom that has lost or gained
one or more electrons to become
positively or negatively charged.

Isotope atoms of an element with
the same atomic number (number of
protons) but with different numbers
of neutrons and therefore different
masses.

Kinetic energy the energy of motion
that particles have, which determines
the speed and force of their motion.

Mass number the combined total of
protons and neutrons in the nucleus
of an atom.

Organic chemistry the chemistry of
carbon compounds.

Oxidation the loss of electrons; a
chemical reaction where electrons
are lost, as when reacting with
oxygen.

pH scale a scale of acidity/alkalinity
expressing concentration of H+ ions.
The scale is logarithmic, meaning
that an increase or decrease of a
value changes the concentration
by tenfold.

Pneumatic chemistry the study
of gases.

Radioactivity the decomposition
or decay of an unstable nucleus,
involving the loss and/or
transformation of subatomic
particles with release of energy.

Valency the combining power of an
atom, ion, or radical; the number of
hydrogen atoms it can bond with.